The Science Myth

God, society, the self and
what we will never know

The Science Myth

God, society, the self and what we will never know

Dominique Chu

BOOKS

Winchester, UK
Washington, USA

First published by iff Books, 2013
iff Books is an imprint of John Hunt Publishing Ltd., Laurel House, Station Approach,
Alresford, Hants, SO24 9JH, UK
office1@jhpbooks.net
www.johnhuntpublishing.com
www.iff-books.com

For distributor details and how to order please visit the 'Ordering' section on our website.

ISBN: 978 1 78279 047 1

A CIP catalogue record for this book is available from the British Library.

Design: Stuart Davies

Printed in the USA by Edwards Brothers Malloy

CONTENTS

Dedication

For XWW

Acknowledgements

There are a number of people without whose emotional, practical and intellectual support this book would not have become a reality. Most of all I thank my beloved wife XS for giving me peace, tolerating my moods and providing me with all that I need to write.

Many ideas that have gone into this book have been developed during my numerous stays at the Senter for Vitskapsteori, University of Bergen, Norway. I thank all the people there and in particular the participants of the Reflexive Systems Biology project for providing such a stimulating intellectual environment. In particular, I thank Prof. Ragnar Fjelland, Prof. Roger Strand and Dr. Dorothy Dankel for many inspiring and often quite heated discussions. These exchanges shaped and continue to shape my thinking. Lastly, I thank the Norwegian Research council for part financial support (contract 187969/O10 Reflexive Systems Biology).

Part I

Chapter 1

Introduction

Humpty Dumpty sat on a wall,
Humpty Dumpty had a great fall.
All the king's horses and all the king's men
Couldn't put Humpty together again.
English nursery rhyme

I remember very well when I met my biologist collaborator on that day. His face was beaming with excitement and I knew that something really wonderful had happened. Probably one of his experiments went very well and he had new results. Maybe, something that would enhance the scientific article we were writing and allow us to publish in a higher rated journal. Unexpected things are often good in science. I was getting excited too.

"So," I opened the meeting, "I guess you have some new results?!" He is not the kind of person who likes to beat around the bush, so he came straight to the point. Time in academia is precious and conversations are kept to the efficient minimum. "Indeed," he said, "we do have new and exciting results." I shifted a bit on my chair, preparing myself for the news. He continued: "We just got the results back from our clinical strains and it seems that they behave the same way as our lab strains."

At this point I felt a sense of disappointment. My collaborator is an expert on *E.coli* and his research, at least he claims, is relevant for the understanding of a number of human diseases. He spends most of his time trying to work out the details of precisely how these bacteria work. To the funding agencies who give him hundreds of thousands of pounds he always makes it clear that his science is relevant for human health. Knowing this

I never expected that there may even be any doubt that the clinical strains, that is, bacteria that are isolated from actual patients, show the same behaviour as the laboratory strains that he uses for his experiments. With hindsight, I acknowledge, this assumption was naive. I should have known better. For several reasons.

I am a trained physicist. And one of the first lessons you learn as a physicist is that the real world does not reveal its secrets. You learn about nature by considering ideal gasses, mass points that have no extension, and motion in worlds without friction. I spent my undergraduate years all five and a half of them at the Department of Physics at the University of Vienna. If there was something I took away from these years, then it was an appreciation of idealisation in physics. Thinking in terms of simple models became second nature to me.

What I did not understand until that day, when I met my collaborator, was that other sciences are probably not very different. Yes, I knew that engineers tend to take short-cuts and do whatever works. I also knew that biologists have to somehow simplify their systems. Still, I never realised what the nature of these idealisations in biological sciences was. Typical microbiologists who work with, say, the bacterium *E.coli* or the fungus *Saccharomyces cerevisiae* will not collect their bacteria or fungi from their gardens, nor do they go out into the forest to get them. Instead, they use domesticated varieties. At some point long in the past an *E.coli* cell has been isolated, maybe from a patient, maybe from a researcher. From then on it has been kept in jars and vessels, fed with pure sugars, multiplied and decimated and passed on from one generation of biologists to the next. In the process, the bacterium has adapted to the lab environment. It has changed its genetics. Maybe it has shed genes that are needed in the wild, but are not that important in the rather protected environment of the lab. In any case, the original *Saccharomyces cerevisiae* or *E.coli* cell has undergone evolution during its travels through the labs of the world. It has become a different system from what it was. It has become a model, a living model, but a model nonetheless. Not very different to the simplified models that they taught me about in Vienna.

Once I understood that the models of biologists are their lab organisms, it all made sense. My collaborator was excited because his model proved to be correct. He probably suspected all along that what he observed in his bacterial pets was relevant in the wild, but he could not be certain before he had actually tested this. This begs of course one question: If the biological model is potentially irrelevant, why did he not do the research on the real thing, the clinical strain, to begin with. The answer is simple: Sometimes the biologist will do just that, but in most cases it makes little sense. There is no single unique clinical strain. There are many variants out there. Even if the biologist used a particular clinical strain, what he finds may well be irrelevant to other strains. Moreover, the clinical strain would probably quickly start to adapt again to the laboratory environment. Before too long it would be domesticated. So nothing is gained. Yet one thing is lost: Reproducibility. When my collaborator does an experiment on his particular strain of *E.coli* everybody could repeat these experiments because everybody has access to the lab strains. If he used a particular clinical isolate, then this would be much harder. Anybody who wanted to do the same experiment would have to ask him for a sample of his strain. Altogether an impractical thing to do.

This is a general lesson about science, a lesson that is all too often forgotten. Science is not about nature. Science is about domesticated strains of nature. This is true in physics, as much as it is true in biology, chemistry, climate science, economics and any other type of research topic. If it is not domesticated, it cannot be dealt with.

Here is an amazing fact about science: It works. At least it works in many cases. The research my collaborator conducts will actually help us understand various diseases. He will probably not come up with a cure, but if someone does one day, it will be because of people like my collaborator. More than that. Look around you and try to think away all the things that are ultimately based on science. The mobile phone will disappear, as will your computer, your TV, your radio, aeroplanes will vanish from the sky, cars, most drugs, and medical treatments. Then, if we thought away science, it would not only affect the material world. Our intellectual lives would be hit just as badly. How much do we understand now about the world that we did not understand before science and how has this understanding influenced the way we think about ourselves, the way we behave and the way we organise society? It is not that long ago that people did not even know what was beyond the horizon, that they did not know what electricity is and they had never heard of dinosaurs. Thanks to science, our internal worlds have expanded in tune with our knowledge of the external world. We are different people now from our pre-scientific forbears.

Science continues to have impact on our lives. We know much and we learn even more at faster rates. New diseases get cured, novel phenomena are discovered, and exciting devices are invented. This is all thanks to science. The best is that there is no limit in sight. There may be many things we do not know today, but surely, at some point we will find out. All this makes science more than just a source of knowledge. It is also a source of hope, a world-picture or *weltanschauung* as the Germans would say.

This is now also where the problem starts. When science becomes a source of hope it really has moved beyond its original remit and encroaches into the core business of other human systems. Most of all religion. At this point, vested interests are affected and conflict seems inevitable.

One direct consequence of this encroachment is the culture war that is going on now. Religion versus science. At first sight, this culture war is not that obvious. Many people live their lives happily without getting in-between the front-lines. However, scratch a bit under the surface of popular culture and the battlegrounds become quickly apparent. I am not even referring to evolutionary biologists turned priests of atheism or evangelical Christians who seem to have an inbuilt scepticism of science. I mean the more subtle symptoms of this raging conflict. The rifts that open up within ourselves, the intellectual choices we think we have to make, and often make. The man on the street who has an intuitive sense of the tension between religious beliefs and scientific thinking and feels that he has to somehow take a position. It can be quite revealing to ask an unsuspecting member of the public what she thinks about science and religion, whether there is a God or whether Adam and Eve existed. Invariably, you will get a response that involves some sort of reference to scientific statements contained in the bible (or any other sacred text). Probably your correspondent will point out that the Christian cosmogony where a creator God shapes the world in six days is not compatible with the insights we have from physics; hence Genesis cannot be true.

Yet, I claim that the conflict between religious creation myths and other apparently scientific statements contained in sacred texts and the results of modern sciences is not the real reason for the tension between religion and science. The cause of the culture war is not that science slowly supplants the need for religion. The real cause is that science becomes more like a religion and religion more like science.

The underlying reason for the culture war is that science is not just facts, it is an integral part of society, of our life and has taken over some of the roles that are traditionally associated with religion. At the same time, science has become pervasive and aspects of scientific modes of reasoning have entered popular thinking. We all think a little bit like scientists; certainly a bit more than our forebears did. As a consequence, we find it much harder to distinguish meaning from truth. For us denizens of a scientific age it is difficult to read Genesis and not take it either at face value or to dismiss it as "just a story." We are not good any more at decoding symbolic accounts. This is largely the influence of science on our lives and it is an idiosyncrasy of our times. The people before us had a much more relaxed relationship to myth and ritual.

Ultimately, the culture war is based on a misunderstanding. The two sides in this conflict are at war because they have become too similar. The scientists have constructed their own piece of religion and the religious have started to think like scientists. This is why they feel they are encroaching on each other's territory. The war is based on a misunderstanding of science and a misunderstanding of religion.

In this book, I will not be interested in religion, but I am interested in science. I hope to be able to convince you that many things we associate with science are myths. The idea that science is rational—myth. Science is value free—myth. Scientists use *the scientific method*—myth. Science is the only way to create useful knowledge—myth. The idea of science as this grand force that marches forward blowing the trumpets of progress does not exist.

Make no mistake. I too think that science is impressive. I love it. I am a scientist. The theories and ideas the great thinkers have come up with are not only useful, but (at least in my view) they are paragons of intellectual beauty. I feel immensely privileged to be able to spend my life participating in this great endeavour. My contribution may be humble, but I feel proud to be able to participate in it.

The question is not whether science is a good thing or not. It is a good thing, but it is also misunderstood. Misunderstood by the general public, misunderstood by politicians who fund it and misunderstood by the practitioners themselves. Many bright philosophers of science have spent many academic lives trying to find out what science really is. Needless to say that they have not come to a final conclusion. Philosophers never do. However, there is some consensus emerging about what science is not. First of all, science is not a unified activity. Scientists do not follow a rule-book. There is no such book. Scientists do whatever is necessary and whatever is possible. What counts as right and wrong and what counts as a proof is different for my collaborator, for a mathematician, a climate scientist or a physicist. Science as an activity looks different in all those fields.

It is often argued that scientific theories are independent of culture, race, political views, religious beliefs and ethical position. The theory of relativity is either true for everybody or nobody. Physics is the same in Russia, Burma, Ecuador and San Marino. An *E.coli* cell functions in exactly the same way independent of whether the scientist is a Pastafarian, Buddist or evangelical Christian. The values of the scientists cannot and should not matter for the results of the research.

As long as we mean the outcomes of particular experiments of calculations, then science is indeed value-free. However, there is more to science. For one, science has consequences. Its products propagate out into the world at large and change the way we think, including the way we make ethical choices. This becomes clear in problems of large scale decision making. "Should we build new nuclear power-stations or leave nuclear energy for good?" or: "Should we allow genetically modified foods, or not?" It is always tempting to settle those disputes using scientific arguments alone. However, this would be a bad idea for two reasons: Firstly, when science is applied to real world problems then there are irreducible uncertainties. Many of the mechanisms that make laboratory science robust are no longer available. Secondly, values become important. Whether or not a nuclear power plant should be built cannot be boiled down to facts, but also depends on how citizens feel about this technology and the risks associated with it.

Sometimes there is a tendency amongst scientists to discount the importance of value choices. After all, so the argument goes, science is based on facts and value choices are just that: arbitrary and subjective choices. It is often overlooked that it is completely unclear why science should be afforded the privileged position it has. Why should a scientific argument be any better than a religious one, or an argument based on values? What is the precise reason why we think that scientific knowledge is superior to intuitive knowledge or knowledge derived by divine revelation (or indeed any other type of revelation)? Why are we so confident that the modus operandi of science leads to results that are better than other activities in society?

At a first glance, it seems obvious why science is superior to other types of knowledge. However, on second thought the problem reveals itself to be harder than anticipated. In the chapters that follow it will become clear that it is difficult to argue rationally why science is any more reliable than non-science. Indeed, on purely logical grounds the argument cannot be made. There is no rational argument that science produces anything better than the teachings of the Upanishads.

It gets even worse. It is not even clear what science is in the first place. There is no definition of science. Philosophers have now tried for centuries to characterise it, to distinguish between activities that are science and that are not science. No luck. The reader may think that this is a non-problem. We all recognise a scientist when we see her. This should be enough, but it is not. If we really think that, for whatever reason, science is superior to non-science, then we should at least be able to say when somebody has stopped being scientific. Yet, this is not possible. The consequences are far-reaching and not confined to academia. For example, if we ask experts to tell us whether or not it is safe to eat genetically modified food or to build a nuclear power station, how can we be sure that their conclusions are based on science and not something else in which case they would be just as irrational as we are. As long as we do not have a criterion for science, we cannot tell that their conclusions are any better than Voodoo.

When it is unclear what science is in the first place, another problem is that science itself is not equally suitable for all kinds of activities. This leads us back to my collaborator. Scientists do not study natural systems. These would be too difficult to understand. Instead, they study domesticated systems. As a scientist you do whatever you can to get some understanding. And this always involves replacing your real thing with some imagined thing that is similar, in some way, but much simpler to understand. Surprisingly, this strategy works extremely well in many cases. It has its limitations though. There is only so much you can understand about wolves when all you study is a Cavalier King Charles Spaniel.

Science is great when it comes to basic physics, chemistry and some biology. However, it is miserable when dealing with complexity. It is chronically bad at predicting not only the behaviour of the most complex systems, but indeed of all but the very simplest systems. When it comes to understanding, it is not much better. Science is at its best in the laboratory where the researcher can control the conditions exactly and do as many measurements as is feasible. The moment she leaves the laboratory things get much more difficult and as a consequence the ability to understand, explain and predict goes down rapidly.

There is no shortage of optimism that science will overcome this limitation. And who knows. One day science may indeed be good at dealing with real complex systems. However, it may equally not be able to get there. All that matters now is that we do have not a shred of evidence that there will ever be a science of complexity. What is true in general, is true for specific problems: Whatever open problem there is in science, it is not rational for us to believe that this open problem will be solved one day unless we have a crystal ball that can truly predict the future.

I do not think that there is anybody with a magic crystal ball. Still, the belief in eternal scientific progress is widespread. At least tacitly it is invoked all the time particularly in popular literature. The form of the argument is always the same: It starts with the realisation that some phenomenon that has previously been unknown is now understood a little bit. For example, modern neuroscience can now measure some aspects of consciousness. During the second step all the unsolved and puzzling questions are listed, open problems are formulated and research programs and manifestos are written. During the third step the faith in eternal progress is invoked and it is predicted that **w**ithin **f**ive **y**ears all the problems will be solved. Sometimes the time span is twenty years or ten years, but it is never 3 weeks or one year. This is what I call a WFY argument. When WFY arguments are made, then this is usually a bad sign for the field in question because there are always unexpected difficulties and five/ten/twenty years later little substantial progress has been made relative to the declared goal. This never prevents anybody from making another WFY argument.

Unquestioned and naive faith in science is widespread. We never consider that science has an appalling track record when it comes to solving problems connected to complex systems. This is not just a theoretical oversight, but it can have practical consequences. Governments typically use scientific input to make decisions in the context of complex systems. Medical treatments are interventions into complex systems (human beings), and these interventions are ultimately based on science. Business leaders and captains of industry have to tackle complexity in their daily dealings. A better understanding of what science actually is, when it is good and when it works not so well, can be useful to help us overcome some naivety and make better decisions.

This optimism is also relevant for the view of science within our culture. Faith in the eternal progress of science and its ability to reduce everything to a naturalistic explanation has consequences for religion. If there is indeed nothing that science will not explain one day, then the Gods have to be reduced to mathematical formulas; better they have to be abolished. The same is true for the self, the soul, the essential "you" which also will be explained and ultimately disappears during the naturalistic turn.

This faith in scientific progress must clash with religion. There is therefore a connection between science and atheism. I am not convinced that atheists are on average more knowledgeable about science than non-atheists, or even believers. There certainly are many examples of eminent scientists who do believe in a God and many scientists who are agnostics and do not care. However, among those who are atheists, I bet there are many who would cite science as a reason for their atheism.

Over the next few chapters of this book, I hope to convince you that these atheists are wrong. Not wrong in their disbelief of the existence of a God. This is up to them. What is wrong is their insistence that scientific thinking and an understanding of science compels us, or indeed compels them to become atheists. I will argue that this conviction is based on a misunderstanding of science. Science is not a sacred and special truth machine, but a social activity that produces useful knowledge and insights in certain areas, but fails in others. Science is not in any way more special than hair-dressing, carpentry or rugby-playing and consequently cannot say more about the existence of Gods than those activities. Nobody would argue that Gods do not exist because of the outcome of the last rugby match.

This book is not a book that tries to take a specific side in the current atheism debate. Indeed, the book is not meant as a direct contribution to this discussion. I do not wish to become part of the culture war. It seems to me that both sides in this war suffer from tremendous irrationality. Arguments are not made based on careful weighing of arguments, but they are all too often emotional, misinformed and poorly reflected. This present text was written with the express aim of providing a more considered input to the current debate, however, without itself taking sides.

Outline of the book

This book is divided into two parts. The first part is about the foundations of science. Its aim is to illustrate the way scientists work by looking in detail at specific examples in physics, biology and complexity theory. Unlike many other authors of popular science books, I try my best not to patronise the reader. I am not simply presenting facts, as is so often done, but I hope that my exposition will provide the reader with an appreciation of the thinking that led to the specific theories. In these first three chapters I am interested in highlighting the scientific achievement as much as the basic assumptions on which the science is built. This is not a textbook. Still, it was necessary to explain some key-concepts of science which may in places make the text somewhat difficult. Chapter 3 which is about Jacques Monod's invention of biology is conceptually the most difficult one. I ask the reader for patience.

The subsequent chapters in part 2 of the book are more accessible and make easier reading. While the first part tries to explain science from an "internal" perspective by showing how scientists come up with their theories, the second part takes an external point of view. In this second part I am interested in the effect of science on society, and on our thinking. I try to find out why we believe that science is a useful activity, what science is and whether it is rational to think that scientific knowledge is in some way better or truer than other types of knowledge. Most of all, I consider whether there even are any other types of knowledge.

Towards the end of the book I start to focus on neuroscience and how it defines, or appears to define, the relationship between the subjective and the objective. I will make the argument that our subjective experience is an example of a phenomenon that is both real and tangible, but *in principle* is beyond science. The inability of science to ever explain this phenomenon is not due to some practical limits to research, but due to the very nature of science.

Chapter 2

Ideal worlds

Weia! Waga!
Woge, du Welle,
walle zur Wiege!
Wagala weia!
Wallala, weiala weia!
Das Rheingold, Scene I

As an academic who is now well into middle age and his first midlife crisis it is increasingly hard for me to impress the young women of this world. Not only are my looks no longer as fresh as they used to be, the topics I like to discuss have little appeal to the ladies. In my desperate attempts to get their attention and interest I like to mention that I am in the modelling business, which is the truth. The problem is that they quickly lose the little bit of interest this may have sparked when they find out that I am modelling bio-molecular systems. Back to square one.

In my view, the young women ought to be interested in my work at least as much as they are in the catwalks of this world. If anything, scientific modelling is more interesting than a career as a semi-starved clothes-hanger who can only eat salad leaves for lunch.

In truth, modelling as in fashion and modelling as in research paper do indeed share some fundamental similarities. The similarity that interests me most is that of idealisation. The models we see in high gloss magazines are not real women, at least they are not truthful representations of the women they depict. The raw image taken by the photographer is always heavily modified, airbrushed and retouched before it becomes the glamourous advertisement we see. Even the near perfect super-models have their little flaws that need to be hidden. Skin impurities are removed, the waistline is made smaller, the skin tone is enhanced, the teeth are whitened and so on. Models are idealisations of real women, their high-gloss magazine pictures are even further idealisations of the models. They are creations of our culture that project the particular idea of beauty that we have right now. They are not about realism.

In science, at least in physics, it is exactly the same. The models that physicists make of the real world are not representations of the world as it is. They are idealisations of the real world, intellectual constructs that formulate how things should be, rather than how things really are. Just as the women from glossy-magazines are non-existent, so are the idealised physical systems of textbooks. They are figments of our imaginations. That fashion models are highly idealised will come as no surprise the reader. As media savvy consumer we know that. However, scientific models as well?! Have they not told us that science is a truthful depiction of reality rather than a fabrication of the human mind?

Idealisation has been shown to be an immensely successful strategy in sciences, in the sense that they have been able to fuel the socio-techno-economic growth seen over the last 400 years or so. At the same time, this enormous success is also surprising. The process of idealisation strips away most of the things that complicate the analysis of a system, but seems to preserve a relevant core. What is left after the idealisation, is both tractable and contains some fundamental essence of the system. Today, the process of idealisation has become second nature to scientists and an accepted approach to generating knowledge. This consensus about the usefulness of idealisation as a method makes it easy to forget that as a concept it is far from easy. At some point somebody had to come up with the idea to consider idealised systems, rather than real ones. The achievement of this invention is bigger than it may seem.

All bodies fall at the same speed

One of the first lessons in physics will invariably be about the law of falling bodies: "All bodies fall at the same speed." But, it is a simple empirically verifiable fact that bodies do not fall at the same speed—that is unless we assume idealised conditions. Like nearly all laws and theorems of physics, the applicability of the law of falling bodies is limited to a number of implicitly made assumptions and provisos. One of them stipulates that it will only be true if there is no air friction. There are others. Common to most idealising conditions is that they are practically never fulfilled in real systems. This begs the question as to why one would want to bother with idealisation in the first place? Why base a science on fictional assumptions that bear little resemblance to experienced realities? The simple answer to this, as will become clear below, is as follows: There is little physicists can say about the world without idealisation. If there are to be tractable theories and laws about nature then they must be about ideal systems. What is even more important though: Idealisation works. Idealised systems, even though they are practically never realised, are relevant to the understanding of real systems. Many of the theories whose scope is strictly limited to a set of simplifying conditions are approximately valid and very useful even under real conditions. The dirty reality we experience can often be considered as just a perturbation to the ideal conditions that make physical laws exactly true. In a nutshell, this is why physics works.

The idea of idealisation and idealised systems is deeply ingrained, not only in physics but in many sciences, from economics to biology. An intuitive understanding of the principle of idealisation has also percolated into wider society. Today, idealisation and simplifications in various forms have become the staple of the scientific *modus operandi*, to the extent that it is hard to see how one could do science without them. Yet, self-evident as it may seem now, the idea of simplification had to be invented at some point.

The inventor is Galileo Galilei, the Italian physicist, engineer and genius who is often seen as one of the fathers of modern western science. He was not only a great scientist, and astronomer, but also a technologist who built telescopes and compasses at the bleeding edge of technology. What he is most famous for today, however, is his systematic study and description of the law of falling bodies.

At the time of Galilei the received wisdom was that bodies do not fall at the same speed, but instead that the speed of their fall depends on their weight. Galilei dispensed with this theory and formulated instead his own law describing the motion of objects and founded modern physics.

There were three key-innovations in his new physics that ultimately would revolutionise science, turn it from the pastime of philosophers to a powerful economic and social driving force that would completely transform society. The first one was the aforementioned method of idealisation. Galilei's law that all bodies fall at the same speed, is not true in real life, but only under idealised conditions. If the idealisation is not fulfilled, he acknowledged, bodies do not follow this law. This is not because the law itself is not valid, but because real life situations have additional effects that modulate the pattern of motion.

A falling leaf in the autumn breeze will slowly glide back and forth, driven by the wind maybe even be taken up in the air again before eventually landing, possibly quite far away from its tree. A poet may find this interesting, but a scientist who wants concise descriptions and general laws cannot do much with falling leaves and breezes. The motion is not only too complicated to be described, but it will also depend on so many other aspects, such as the weather, and the shape of the leaf that it would be quite useless in practice. Even if there was a description that could take care of all these aspects, it would be very specific to the particular case and lack the generality that science is interested in. Even if we knew the behaviour of a leaf, it tells us little about how a stone falls.

Now, idealise the system, take away air-friction and the problem simplifies significantly. Not only that; it also becomes more general in scope. The impact of this first of Galilei's methodological innovations can hardly be underestimated, both in terms of its importance for science but also in terms of its conceptual profundity. Once air resistance is ignored, the difference between a falling leaf and a stone and indeed any other falling object disappears. They all fall at the same speed. One key idealisation gave Galilei the license to neglect nearly everything about the individual falling objects.

For Galilei the principle of idealisation prepared the ground for a second key innovation. Once he had established that all bodies fall at the same speed, a question came naturally: What is this speed? How fast are bodies falling? Galilei answered this question with his square law:

The distance travelled by a freely falling body is proportional to the square of the time since it was dropped.

Formally, the same statement can be written as an equation in the following way:

$$d = Ct^2$$

Here the letter d represents the distance travelled by the falling body, that is the dropping height. The symbol t is usually used to denote time; conventionally the time is assumed to be $t = 0$ at the moment when the object is dropped and then continues onwards in, say, seconds. The symbol C is a constant number which is unknown at first and needs to be determined by experiment. This formula can now be used by anybody to predict precisely how far any object, independent of size and shape will fall in a given time and hence how fast it falls.

If the above formula is to be used in practice, then the letters d, C and t must be replaced by concrete numbers. It has no poetic appeal, and we do not know whether the falling leaves are red or yellow, and there is no mention of the tree. Indeed, even if we have all information about the falling body, we will not be able to input this into the formula. By throwing out all these details Galilei managed to make statements about the behaviour of natural objects, such that it is convenient to reason about them quantitatively.

The second of Galilei's innovations pertains to his square law. The point is that it describes nature using mathematics. This introduced a fundamentally new kind of description. Rather than merely making qualitative statements about the world, formulas and mathematical reasoning made it possible to give *quantitative* descriptions of natural processes. This is the difference between saying that it is really hot today and specifying that the temperature is 34° C (which some may not find that hot). The first is a qualitative statement and the second is a quantitative one. Much of pre-Galileian science was merely qualitative. Numbers and quantities are the bread and butter of modern science.

The third of Galilei's innovations is closely connected to the idea of a quantitative description of nature: Experimental tests. The square law makes a prediction about how falling bodies behave, in the sense that given a distance d (which is in essence the height) the law predicts the time t it takes for any body to fall this distance. The only way to find out whether or not the square law is correct is to compare the predictions of the law with the behaviour of a real system. The basic idea is very simple: First, build an experimental system that realises the conditions under which the law is valid. Then use the law to make a prediction about the outcomes of the experiment. Finally, compare the results of the experiment with the predictions. If the experiment agrees with the prediction in a sufficient number of cases, then it is reasonable to assume that the law is correct. Otherwise, the law probably needs revising.

In practice the situation is much more complicated than this. One of the main difficulties of experimental science is hidden in the formulation "...experimental system that *realises the conditions under which the law is valid*." The square law is only correct in an idealised world. This means that the law itself also must be tested under these idealised conditions. In the case of the free fall, one way to do this is to create a vacuum and perform drop experiments in it. This seems to be an attractive option at first sight, but in practice it is not feasible, at least it was not feasible at the time of Galilei.

Idealisation in this sense could have developed into a sticking point for the nascent Galileian approach to science. By restricting his theories to idealised systems, Galilei was able to postulate general and quantitative laws. The downside of this is that these simple laws are not applicable to real world systems and are untestable unless the idealised conditions can be approximated. This dilemma could have easily spelled the end the new science before it even began—had Galilei not found an ingenious way around it by inventing the *laboratory system*.

He knew that he would not be able to realise the idealising conditions to test his hypothesis about the free fall directly because he was unable to create an effective vacuum. What he could do, however, was to try to approximate the ideal conditions as closely as possible or to design experiments where the air-friction can be systematically factored out. There are many possibilities to do this. One is to drop objects of different mass, but same shape, size and material. As long as both objects are heavy enough, one would expect that they are equally affected by air friction and should therefore fall at the same speed; even under non-ideal conditions. Another possible experiment is to place a piece of paper at the back of a book and drop the book. The book will shield the piece of paper from air resistance and consequently both will fall at the same speed. The design of the experiment has factored out air resistance. A variant of the experiment is to drop paper and the book side by side. In this case the piece of paper and the book will be affected differently by the air resistance; the former will slowly glide to the ground whereas the book will drop just as in the first experiment.

These experiments can be easily performed without creating a vacuum. They do not directly demonstrate that all bodies fall at the same speed in vacuum, but they do demonstrate that air resistance alters the motion of falling bodies in a way that depends on their shape. By implication one could hypothesise that in vacuum bodies fall at the same speed.

Simply showing that all bodies fall at the same speed was only part of Galilei's agenda. A much more important aspect was to experimentally confirm his square law. This required quantitative data about the time it takes for objects to fall a certain distance. This quantitative information cannot be extracted easily from the two suggested experiments above, but required more elaborate systems.

There are many challenges in the way of performing accurate experiments. One problem concerns the requirement to control the experimental error. In the case of Galilei this meant that he had to perform a large number of measurements to confirm his square law. Naively, one may assume that it is enough to make a measurement and check whether it agrees with the prediction. In the real world things are never that easy. Just one of Galilei's problems is with the unknown constant C in his formula. In essence, what this constant does is to set the dimensions of the other quantities; that is to say whether distance is measured in inches or meters or yards or whatever. The same for time. This constant cannot be derived from theory, but must be measured. Galilei could not do that. Therefore Galilei could not predict the precise time of a fall but only how the time of the fall changed as the height was varied. So, in order to confirm his square law he had to ascertain the shape of a curve rather than measure absolute timings. However, he only needed to do this once to determine the value of C. Once this value is known, he could calculate exact falling times.

In order to confirm the shape of a curve one needs at least three points. One point is clearly not enough. A single data point gives no information about the shape of a curve. Similarly, two points can always be connected by a straight line or indeed fit any shape. Three points are therefore the minimum number of measurements, but in reality Galilei needed many more. One of the main problems he had to contend with was experimental error. Realistic measurements always have a degree of uncertainty associated with them. This introduces some ambiguity about the shape of the curve on which the points lie. So, in reality, when physicists measure a quantity, then these measurements will normally not directly coincide with the expected value, but will be a little bit different. In practice this manifests itself graphically in that several measurement results do not lie on the theoretically predicted curve, but a bit below or above. Figure 3.1 illustrates this schematically with some artificial data. The top panel shows an example with many data points and a relatively small spread of the data points (that is low error). From those the square law (indicated by the solid line) can be inferred with good confidence. This contrasts with the bottom panel that shows a case with few data points and a wide spread. The experimental values may well fit a square law, but it is probably also compatible with other curves. So, for example, the three points look as though they could be fitted to a straight line. Maybe one can also fit them with a cube law, or one of many other possible shapes. The point is that if just these three points are given, it would be very difficult to guess that they belong to the square law that is indicated by the solid curve in the graph. By extension, if Galilei only had those three points he could not be confident to have demonstrated the square law. The result would be inconclusive.

For Galilei this meant that he had to be able to do two things if he wanted to demonstrate his square law. Firstly, he needed to be able to make many different independent measurements of falling objects, so as to generate many different data-points. Secondly, for each of these measurements he had to reduce the error of measurement sufficiently so that he could be confident that his sample points lie on a curve consistent with the square law.

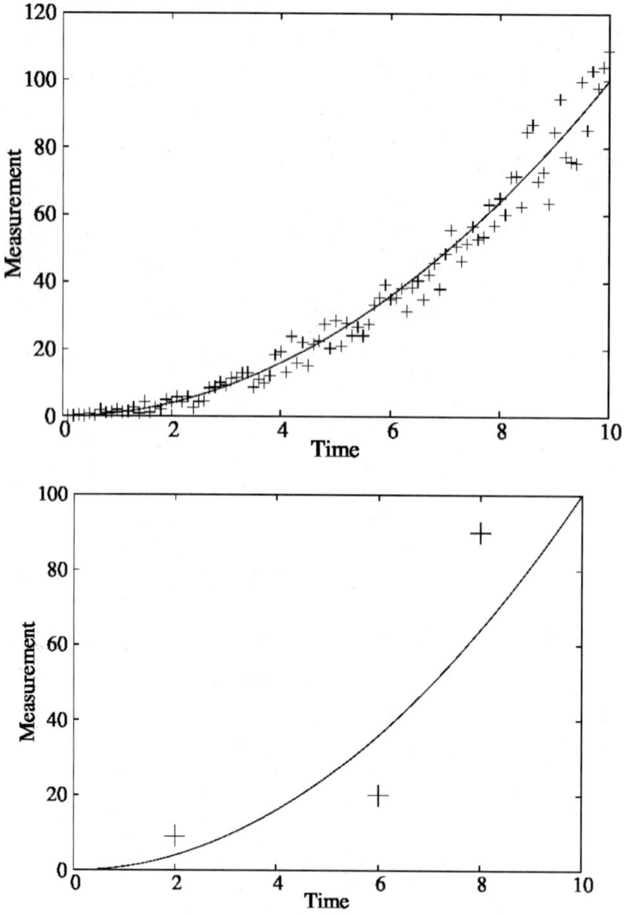

Figure 2.1: Illustrating error in a hypothetical experiment. The crosses could be measurement results and the line is the theoretical square law. The top graph has many measurements with low error. The bottom one few measurements with high error; it will not be a convincing confirmation of the square law.

Galilei's experiment

Naively one may think that measuring the time it takes for an object to drop is a trivial thing to do. It is not! At least not when your aim is to prove that falling objects follow a square law and you only have the technological means of the 16th century at your disposition. Galilei knew that this was not going to be an easy task. It would not be sufficient to pick up a stone on the beach and drop it a few times. Some serious consideration and thought was required.

The first problem Galilei faced was to approximate the idealised conditions. The square law is only true in the absence of friction, air-friction and other types. So somehow, he had to eliminate this. There are two ways to do this. Either Galilei could have tried to remove friction altogether or, which is much easier, find a setup where friction can be neglected or factored out at least.

The other challenge he had to meet concerned the number of measurements he needed. He knew that he would not be able to measure falling times precisely, so he needed many data points, a large number of independent measurements of the free fall. So, any arrangement he would come up with had to be such that Galilei could easily vary the height of the fall to be able to get as many data points as possible.

The third crucial aspect he had to get right concerned the setup of the experiment. Galilei knew that he had to measure the time of the fall as accurately as possible. Precisely how accurate depended on the number of experiments, or the number of measurements he could perform. A high number was important because more points, particularly if measured over a wider range of heights, increased the confidence in the result. Apart from the raw number of experiments, the range of conditions over which he could do experiments was also important. In his case the range was in essence the maximum height from which he could drop objects. The range is important because it determines how close individual measured points have to be together. The closer points are together, the harder it is to distinguish the hypothesised square law from a possible linear law or a cube law or indeed any other type of law. So, if the range is small, then Galilei would have had to compensate with a higher precision of individual measurements.

Legend has it that Galilei used the leaning tower of Pisa in Italy to test his square law. He was indeed born in Pisa and certainly knew the tower but there is no historical record of him actually using the tower to perform his experiments. The famous landmark would have made a beautiful backdrop for this important moment in the history of science, but what it had in romantic appeal, it lacked in suitability. Using the tower of Pisa to confirm the square law would have been impractical. It is high and walking up is strenuous. The experiments would have been outdoors which would have put him at the whim of the elements. It could rain at any time, winds could disturb the drop experiments, in the summer heat it could be even worse to climb the tower and members of the public may disturb the experiments at any time. Apart from these impracticalities, the tower also offers only limited scope to systematically vary the dropping height and to obtain enough data points. It is about 55m high. This is very high when you need to climb it, but not particularly high for drop experiments. From the leaning side he would have probably had 50m of actual dropping height corresponding to a dropping time of 3.2 seconds. Using the technology Galilei had at his disposition measuring such intervals accurately would be difficult, although not impossible. To prove his square law, Galilei needed at least two more points. Halfway down the tower at 25m the drop time is about 2 seconds. It would have been extremely difficult for Galilei to distinguish 2 seconds from 3.2 seconds with an acceptable error, which means that a large number of repeat experiments would have been necessary to average out the error in the timings. For the third point he could have chosen 10m with a falling time of 1.4 seconds. This is probably below the limit of what he could measure. This would leave 35m for a fourth point which corresponds to 2.7s and is difficult to distinguish experimentally from the drop time for 50m and 20m.

Dropping objects from the tower of Pisa was impractical. Galilei needed to generate quantitative data with high accuracy. This is best achieved indoors, in the controlled environment of his laboratory. This is where Galilei went.

Taking experiments indoors created its own problems. If timing the fall of objects from the 50m tower of Pisa was infeasible, it was even more so in his lab where it was difficult to achieve dropping heights of more than a few meters. On the other hand, it is more convenient to time the fall within the laboratory. There would be no climbing involved, lunch could be had just downstairs, no members of the public who disturb, no thunderstorms or heat-waves to bother the scientist. All this allowed for more sophisticated instrumentation to be built and Galilei could hope to achieve higher precision measurements. Despite all these advantages, the big drawback of the laboratory environment is the lack of falling height that Galilei could hope to achieve. For a straight drop he was probably limited to 3m, the space from the ceiling to the floor. Hardly a lot if compared to the 50m tower of Pisa. The increase in convenience and measurement precision would probably not be enough to compensate for the dramatically reduced falling times in the lab. Galilei had to come up with a different idea. And he did.

The main problem when confirming the square law was that the relevant intervals he had to measure were very short, which makes it difficult to achieve high accuracy. One possible solution for this would be to slow down the fall and thus achieve a greater relative precision. Longer times are easier to measure than shorter ones. This would solve the problem if it worked, however, it is not obvious that it does. If Galilei somehow managed to slow down the rate of the fall, he may be able to obtain accurate timings, but perhaps at the expense of making the system irrelevant to the question he was asking. Slowing down the fall manipulates the very quantity Galilei wanted to measure.

This is now where Galilei's genius showed. He found a way to slow down the fall and still remain relevant: Instead of performing a free fall he focused his attention on a different, albeit related, system, namely balls rolling down inclined surfaces. Motion on inclined planes is clearly driven by gravity, so there is a connection to the free fall. At the same time, the setup allowed Galilei to vary systematically by how much he wished to slow down the fall, or in other words, how closely he wished to approximate the free fall. All he needed to do was to adjust the angle of his plane. This is easy to do and can be done quite accurately without too much effort. If the plane is parallel to the ground, the balls would not fall at all; in the case of an angle of 90 degrees, the balls would be in a perfectly free fall. For all angles in-between these two extreme cases, Galilei would have effectively slowed down the fall, enabling him to make accurate measurements. Now, to begin with, he could not be certain that the motion down the inclined planes is the same as the free-fall (and indeed it is not). However, experimentally, he could measure the extent to which the two motions differed. It seemed at least reasonable to assume that the motion becomes more similar to the free fall as the angle increases. So, inclined planes appear to be, at least, relevant to the question at hand.

Once Galilei had decided that the inclined planes were a suitable experimental platform for him to investigate the law of falling bodies, the next problem he needed to overcome was to reduce friction. In the inclined planes experiment he would encounter two types of non-ideal conditions. Firstly, air resistance and secondly friction between the rolling balls and the inclined plane itself. He was not able to remove either of those completely, but could reduce them as much as possible by preparing his inclined plane well and by choosing the right material for his balls. In his "Dialogues Concerning Two New Sciences" Galilei described his experimental setup:

A piece of wooden moulding or scantling, about 12 cubits long, half a cubit wide, and three finger-breadths thick, was taken; on its edge was cut a channel a little more than one finger in breadth; having made this groove very straight, smooth, and polished, and having lined it with parchment, also as smooth and polished as possible, we rolled along it a hard, smooth, and very round bronze ball. Having placed this board in a sloping position, by lifting one end some one or two cubits above the other, we rolled the ball, as I was just saying, along the channel [...].

In this quote from his book, Galilei described in detail how he approximated the ideal conditions that make his square law valid. He carefully prepared his plane by edging grooves into it. These grooves constrained the balls into a straight and controllable trajectory. He polished the grooves and lined them with polished parchment. He even described how he polished his bronze balls. He took all this effort to reduce the influence of friction, not only air-friction, but also friction between the ball and the surface of the inclined plane.

Galilei still had the problem of timing the fall accurately. He did not have access to accurate off-the-shelf timers, as we do today. In order to be able to measure the time of the fall accurately Galilei had to develop his own method in the lab.

How does one measure time accurately in the absence of a timer? One possibility is to use the pulse rate to measure time. The problem with the pulse is that it is not very accurate for short intervals of time. A normal pulse rate is about 40-60 beats per minute, so roughly a beat per second. A very crude reference measure for intervals shorter than, say, 5 seconds. This rules out the pulse for direct drop experiments, but also limits the inclined plane experiments to small angles with low inclinations. Another disadvantage of the pulse is its variability. Physical activity or even just psychological factors may change the pulse-beat. Galilei had to find a novel solution. He described it in his Dialogues:

For the measurement of time, we employed a large vessel of water placed in an elevated position; to the bottom of this vessel was soldered a pipe of small diameter giving a thin jet of water, which we collected in a small glass during the time of each descent, whether for the whole length of the channel or for a part of its length; the water thus collected was weighed, after each descent, on a very accurate balance; the differences and ratios of these weights gave us the differences and ratios of the times, and this with such accuracy that although the operation was repeated many, many times, there was no appreciable discrepancy in the results.

This timer is as simple as it is ingenious. Using a device that guaranteed a constant flow of water, Galilei converted the quantity "time" which is hard to measure into the quantity "weight" which is easy to measure. This simple trick allowed Galilei to make reasonably accurate measurements of time intervals a pre-condition for the experimental confirmation of his square law.

With the experimental design in place, Galilei was now ready to perform the actual experiments. He had to show two things. Firstly, he needed to empirically verify the square law, which was the main task. In addition he also needed to convince any sceptic that his contraptions were in fact fit for this purpose, that his experimental setup could indeed be used to demonstrate the square law. One part of this was to satisfy any critics that his setup was at all relevant for the free fall, when it *prima facie* appears to be about a different system (balls rolling down a surface) whose relationship to the free fall is unclear. Making this case is in essence a theoretical argument that Galilei had to present.

In order to show that his experimental setup is reliable he had to demonstrate that he could control the measurement error.

Galilei had to show that his experimental error was small enough to allow meaningful conclusions to be drawn from the setup. Referring to his experiment on the inclined planes he wrote:

We repeated this experiment more than once in order to measure the time with an accuracy such that the deviation between two observations never exceeded one-tenth of a pulse-beat. Having performed this operation and having assured ourselves of its reliability, we now rolled the ball only one-quarter the length of the channel; and having measured the time of its descent, we found it precisely one-half of the former. Next we tried other distances, comparing the time for the whole length with that for the half, or with that for two-thirds, or three-fourths, or indeed for any fraction; in such experiments, repeated a full hundred times, we always found that the spaces traversed were to each other as the squares of the times, and this was true for all inclinations of the plane, i. e., of the channel, along which we rolled the ball. We also observed that the times of descent, for various inclinations of the plane, bore to one another precisely that ratio [...].

In this passage Galilei described what he considered to be the main result of his experiments. For each angle he confirmed that the rolling ball obeys a square law, and he also showed that his results were reliable, in the sense that repeating them led to the same result each time with only a minimal experimental error.

Note that this did not demonstrate experimentally that free falling bodies obey the square law. In a strict sense, Galilei actually never showed this explicitly. Instead he argued his case, roughly as follows:

1 The square law is fulfilled independently of the angle of inclination of the plane.
2 The free falling body is a special case of a plane that is inclined at 90 degree.
3 Hence, the free falling body obeys the square law.

Galilei's experimental test of the square law rested on a theoretical argument. Technological limitations prevented him from experimentally addressing the free fall directly. Instead, Galilei constructed an elaborate argument that falling bodies should obey the square law, and only confirmed parts of this argument experimentally.

Galilei's legacy

Galilei's square law is still an integral part of secondary school science curricula. Yet, his main revolutionary impact on science is perhaps not so much due to the square law itself, but due to his methodological innovation. He was the first to bring together (i) a mathematical description of nature's laws, (ii) the use of idealised systems to formulate the laws and (iii) the use of laboratory systems that is the use of experimental setups that approximate idealised conditions to check the theory.

From a modern perspective, all this seems, if not trivial, at least somewhat un-spectacular. Physics students are taught to think in terms of idealised systems from their first day at university. By the time a student enters graduate school, this has become second nature to her. Galilei's methods of idealisation, mathematical description and experimental test are now deeply ingrained in modern science, and have even penetrated popular thinking. All this makes it difficult for us contemporaries of a scientific age to appreciate the monumental dimensions of Galilei's achievement. But without a doubt it was this Italian scientist who rang the opening bells of modernity and enlightenment.

Fundamentally it is quite astonishing that idealisation leads to laws that are both simple and general. There is no *a priori* reason to assume that the world is like this, that beneath the layers of apparent complexity a mathematically beautiful reality is hidden. Neither is it clear that we can learn something about the real world by studying simplified worlds only. Indeed one would normally be inclined to think that ignoring effects would lead to less general results, not more general ones. With the benefit of hindsight, it is evident why ignoring friction works and why this leads to simpler systems, but prospectively the route of idealisation must have seemed not only unpromising, but simply absurd and irrelevant.

In science the use of mathematics is often held as a gold standard for the rigour and maturity of a field of investigation. Mathematics is considered to be the language in which God has written the universe, as Galilei expressed it. The usefulness of mathematics to express scientific laws depends very much on the simplicity of the underlying laws and hence idealisation. In idealised systems most properties of falling bodies are irrelevant which makes it possible to formulate equations and mathematical models that are tractable. Applying mathematics directly to the apparent world, the world we experience with all of its complexities is doomed to fail.

Simplicity and mathematical laws in science created their own problems, and a whole new "industry" committed to solving these problems. Since ideal systems are not normally accessible directly, experimental science needs to find techniques to approximate ideal conditions often challenging the limits of what can be done given the means of the time.

Ultimately, the feasibility of Galilei's methodology rests on the world being fundamentally simple. This simplicity is what philosophers call a *contingent fact*, something that *is* the case in our world, but may as well not have been the case. The world could have been very complicated and not amenable to mathematical descriptions at all. The free fall could have depended on the shape of the objects that fall and on the material they are made of. This is not the case, but it could have been. Simply put, something is contingent when we have to check or measure in order to be certain that it is the case. Incidentally, this is the case for most interesting knowledge about the world.

Galilei's main contribution to science was to discover the particular contingent fact that nature, at least some of its aspects, can be described using relatively simple mathematics. He was able to show how this can be used to develop novel and (relatively) certain empirically grounded knowledge about nature. In this way he established a new way of doing science: the Galileian paradigm. This Galileian way of doing science has proved to be immensely successful in two ways. Firstly, the Galileian methodology did not only work in the case of free falling bodies but could be extended to the wide range of phenomena. Yet, importantly, the Galileian science, although fundamentally about an idealised world, was also able to underpin technological development. It did not remain a gentlemen's pastime, a mere natural philosophy, but became a driver of socio-economic change. It was of practical use.

Without taking away from Galilei's genius in any way, we have to assume that he was probably unable to predict the success of his new approach to science. His discovery that the law of falling bodies can be expressed using mathematical means is exciting, but it does not in itself in any way logically allow us to conclude that other aspects of nature are equally simple. Therefore, for Galilei himself and his contemporaries, it could not have been clear that the new science would have the impact it ultimately had. Scientists could have run out of simple laws to discover after the free fall was successfully described. Or the new science may have remained a "natural philosophy" forever, without relevance for real systems, and by extension with no lasting impact on technology and society. Needless to say this is not what happened.

When Galilei invented his new methodology its lasting impact and legacy was probably not his main concern. That idealisation in physics works is an interesting fact about nature. The main reasons that physicists use idealisations are pragmatic ones. Galilei, if he wanted to say something general and non-trivial, had to simplify. There was nothing else to do. Describing the motion of falling leaves in a natural context is simply not feasible. Idealisation offers simple, tractable laws that avoid getting bogged down by details of shape and weight and air flows that determine the fall of bodies under real conditions. By concentrating on the arcane Galilei could actually produce results. The ultimate practical relevance of his law was not the aim of his investigations; and the same is probably true for most of his contemporary colleagues. The technological fall-out from science, its applicability to the real world is predominantly accidental. It could have been different.

Inertial frames

Within physics the use of theoretical and mathematical models and the use of laboratory systems to confirm them has continued to be successful after Galilei. The next milestone was Newton's formulation of his three laws. These three laws are not only of fundamental importance in mechanics, but they also became symbols for the success of science. As far as this narrative is concerned, we will focus on the first of Newton's laws because it makes an interesting case study illustrating how physicists develop theoretical concepts, and how consequent reasoning leads to fundamentally new insights. As such, Newton's laws and the ideas behind it are an illustration of the scientific process itself—at least one type of scientific process.

Of the three laws Newton's second has acquired an iconic status in popular culture that is nearly equal to Einstein's $E = mc^2$ formula. Expressed mathematically it reads as follows:

$$F = ma$$

It is one of the most fundamental laws in classical physics in that it establishes the notion of a force **F** and how it relates to the acceleration **a** and the mass m of a body.

The third law is conceptually the most difficult one. It states that whenever a force **F** is exerted on a body, then this body exerts a counter-force −**F**. What this means is that, if one pushes a wall exerting a force, then the wall pushes back on the pusher with the same force. These counter-forces are quite real and are the physical principle behind, for example, rockets.

In what follows we are going to focus on Newton's first law. *Prima facie* it does not appear to add much to Galilei's insights about the motion of bodies. Indeed, Newton's first law had been known to Galilei in one form:

> **Newton's first law (I):** A body that is resting will not start to move unless a force is exerted.

This can be easily visualised in terms of Galilei's experiments. In the limiting case of a plane that is parallel to the ground the ball will not move at all. This is precisely what we would expect in our everyday lives. A car that is parked on an even surface will not start to move by itself. An orange on the kitchen table will not start to roll onto the floor unless pushed; a basketball will remain unmoved unless somebody picks it up or kicks it. This part of Newton's first law is quite intuitive, and in tune with our daily experience.

There is another, slightly deeper and certainly less intuitive aspect to this first law. To see this, it is best to start from an alternative formulation of Newton's first law:

> **Newton's first law (II):** If a body is at a constant speed, then it will remain at this speed unless some force is applied to it.

As they stand the two versions of Newton's first law seem to be very different statements, although it will become clear below that they are in fact equivalent. While the first formulation of the law is of high familiarity to most of us, the second version of the law we see only rarely fulfilled, if ever. It is easy to find a counter example to this second version of Newton's first law. Think for example of a billiard ball. Newton's law would predict that the ball, if hit, would continue to make its path across the billiard table until the end of all days. It is safe to say that in reality this is not the case. If hit hard enough and if the table is of good build quality, then a ball probably bounces off the boundaries six or seven times before eventually coming to a halt. Newton's first law (II) disproved.

Any physicist reading this will now immediately jump up and object that Newton's law is not applicable to the case of the billiard ball. The table exerts a friction force on the ball which slows down its motion until it stops. Whenever the ball hits the boundary, then it deforms the soft cushion a bit. Some of this deformation will be returned to the ball as the deformation bounces back to its original form like a spring, accelerating the ball in the process. Some of it, however, is irreversibly lost, converted into heat and cannot be retrieved any more. Similarly when the ball rolls over the lining of the table the felt is slightly but continuously compressed, squeezed and deformed, slowing down the ball in the process. Even the best felt will eventually bring the ball to a halt.

These daily experiences of apparent violations of Newton's first law do, of course, not disprove it. It is not that Newton's first law is wrong, it is simply not applicable to the case of the billiard ball, at least not directly. The culprit is, again, friction. Newton's law assumes, just like Galilei's square law, ideal conditions, which amongst other things means that there are no frictional forces in the system.

If one wanted to confirm Newton's law experimentally, then again, one has to build laboratory systems that approximate the ideal conditions. This has been done many times and the law stands confirmed. Apart from direct experimental tests, physicists often use a different type of experiment to explore the implications of laws: *Gedankenexperiments* (derived from the German *Gedanke* meaning "thought") are speculative experiments purely performed in the mind of the physicist to clarify the implications of theories and laws. Such a *gedankenexperiments* will show that there are some deep implications to Newton's first law which will ultimately directly lead to the theory of relativity. Over the next few pages we will perform some *gedankenexperiments* to illustrate how physics works and how physical concepts and theories are derived from one another.

Münchhausen

To understand Newton's first law it is crucial to be clear about the idealising conditions that make this law true. One was the absence of friction. In addition, there is another requirement, namely that any measurements to determine the motion or absence of motion are performed within a so-called *inertial frame of reference* or *inertial frame* for short. An example of an inertial frame is a train or a car that rolls at a constant speed. Put a ball onto a table in front of you, or onto the dashboard and it will remain still, as long as the table is not inclined and as long as the ball is not pushed or somehow actively set into motion. This is in accordance with Newton's first law (I). The moment the car or train accelerates or breaks, it ceases to be a good approximation to an inertial frame. Newton's first law is then no longer valid, and the ball will start to roll forward or backwards depending on whether the vehicle accelerated or decelerated. For any passenger within the train, no obvious force has been applied to the ball and it will look as though the ball was set in motion spontaneously. This does not violate Newton's laws because the accelerating train compartment is not an inertial frame of reference.

Physicists like to think of reference frames as coordinate systems. These are in essence numerical labels for every point in the world. In our 3-dimensional world this could be three numbers corresponding to their position within the reference frame. These days GPS coordinates are very common. They label every point in terms of their latitude, longitude and altitude. One can choose these coordinate system as one wishes. A common choice is to label the points such that the measurement device, or the scientist who performs a measurement, would be at the centre of the coordinate system. Any motion observed by the measurement device is relative to the measurement device and hence relative to the frame of reference. An inertial frame is simply a frame of reference that moves at a constant speed and is not accelerated.

Reference frames are a bit like the cameras that are used to observe sporting events. Nowadays in live broadcasts of Formula 1 races there are two types of cameras. For one there are the standard cameras that capture the race from various positions along the track. This gives an outside view of the race as the spectator at the circuit would experience it. Then there are car mounted cameras. These allow the TV audience to experience the race from the perspective of the drivers, adding to the thrill of the race. When the broadcast switches from the fixed camera to the car-mounted one, then this is a change of reference frame. The race is clearly the same, no matter where the camera is, but the perspective is a different one. In the car mounted camera a particular car has been made the centre of a coordinate system, and is assumed to be "at rest," whereas in the case of a track mounted camera the very same car is moving but the tarmac of the circuit is at rest. Mathematically these two descriptions are equivalent in the sense that one can simply convert one into the other by some simple operations.

Physicists are very fond of this idea of changing between different perspectives, or as they call it in their jargon, coordinate transformations. They love to analyse how theories and statements change as coordinates are transformed as the perspective of an (imagined) camera changed. Converting descriptions from one inertial frame into another, or the viewpoints of non-accelerated "cameras" are a specially important class of transformation. Formula 1 cars would normally not count as inertial frames. They constantly break and accelerate as they negotiate their way around the corners and chicanes on the race course. Although the camera fixed to the track can be seen as an inertial frame (at least to a high degree of approximation).

It may seem at first rather pointless to describe events from different perspectives. It is hard to see what can be gained from it. Over the next few pages, however, it will become clear that a ruthless analysis of events from different perspectives can lead to very deep insights. In what follows we will limit ourselves to inertial frames only and exclude the more complicated case of reference frames that are accelerated. Before we can get into our discussion, we have to ask a preliminary question: How can one be certain about whether any specific frame of reference is accelerated or not, that is whether or not it is an inertial frame?

This question is not as straightforward to answer as one may think at first. One thing to consider is whether or not the objects within the reference frame seem to experience unexplained forces. The ball on the table in an accelerating train experiences such forces, manifesting themselves in an apparently spontaneous motion of the ball. From the perspective of the ball, of course, what happens is that the table on which it rests suddenly starts to move, although to the passenger the motion of the ball appears completely uncaused. For an observer such uncaused motions could indicate that she is not the centre of an inertial frame. Their absence does however not guarantee that she is in an inertial frame.

We will not go deeper into this and circumvent the problem by using Galilei's trick and make idealising assumptions about the kind of world we consider. To simplify our *gedankenexperiments*, we assume that there are no large masses that could exert gravitational pulls. In other words, we imagine to be in a world where there are no internal forces. This imagined world is very unlike the universe we know with all its stars and planets. It is just some dark space, featureless and empty. This is the Newtonian world in which we can investigate the first of his laws and its consequences.

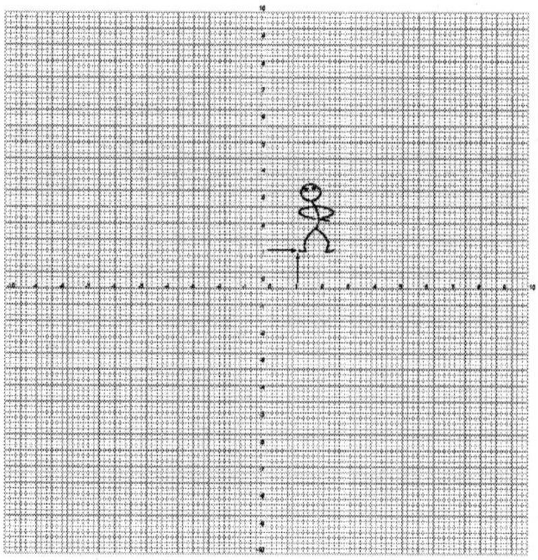

Figure_2.2: A coordinate system in two dimensions. It labels every point in the world. We define the position of the Baron by the location of the tip of his right foot. This happens to be at (1,1) in this instance, as indicated by the arrows.

Within this world, we can now imagine ourselves riding a cannonball like the famous Baron of the Lies, Freiherr von Münchhausen. When laying siege to the stronghold of the enemy Ottoman troops, the Baron wanted to gather some intelligence about the state of the enemy. He tried to find a way to get behind their lines. After agonising over the problem, he eventually came up with the idea to grab one of the cannon balls and ride on it over the fortification of the Turkish troops. Halfway to his goal, he began to worry about how he could get out again unhurt to report his findings. While thinking, he noticed that there probably was no safe way back and decided that it was better to swap cannon balls, grab one of those flying from the stronghold to his own lines and take a ride back.

Baron von Münchhausen's ride on the cannonball makes a good story but is hardly believable. Apart from the difficulty of catching a cannonball as it leaves the cannon, it is quite impossible to sit on a cannonball in midair. One is therefore very tempted to dismiss the ride on the cannonball as just another one of these stories of the Baron of the Lies.

There is another approach to it. Instead of bluntly disregarding the truth of the story, we can start to ask how the Baron knew that he was in fact riding a cannonball, rather than sitting on one that was at rest. It would be much easier to believe that he was sitting on a motionless ball, although this is also much less to brag about. The question about how the Baron knew that he was riding a cannonball at high speed takes us right into the realm of Newton's first law. To make it applicable we need to assume that von Münchhausen was riding his cannonball in a world without gravitation, and without friction, and most of all in a world that had no features other than the Baron himself and his cannonball. We also need to assume that the cannonball the Baron was riding was not subject to accelerating forces.

Given this Newtonian world, we can now challenge von Münchhausen and ask him how he could have possibly known that he was riding a cannonball at high speed, rather than one that was at rest. To defend his claim, he would have to explain how he measured his speed while riding. In the real world, this would be quite straightforward. The Baron would feel the air resistance in his face, and he would be able to estimate his speed relative to the battlefield beneath him simply by observing how the ground seems to pass by underneath. In an ideal Newtonian world without air friction and devoid of any features, Münchhausen would get no such feedback. He would need to find some other way of determining his speed. But how?

If he was in a completely empty world that had nothing in it except the Baron himself, then it would be very difficult for him to be certain about his speed. In order to determine his velocity the Baron would have to have some way to determine his own position and a way to measure time. He could then check how long it takes to travel a certain distance and work out his speed from that. If we grant him the benefit of a time-keeping device then he is left with the problem of determining his position. In the empty Newtonian world, this is not possible, because one part of it looks exactly like any other part of it. The space is a black void and there is nothing in it except for infinite emptiness. There are no landmarks or reference points that Münchhausen could use to determine his speed. With no way to tell where he is, he would also have to remain in doubt about whether he is moving at all or whether he is at rest. Whatever he assumes, there is no way to be certain, so the Baron may well assume that he is the centre of the coordinate system, or the centre of the inertial frame and not moving at all. Yet, equally he could also choose some other coordinate system that travels at some speed relative to himself. In that way he would define himself as travelling at a constant speed. In a nutshell, because the Baron cannot measure his speed whether or not he is in motion or at rest is an arbitrary decision.

Let us now allow a slightly enriched Newtonian world that also contains the cannon from which his ball was fired, as well as the enemy cannons that continuously shoot guns in the opposite direction. In this world there are more possibilities to measure speed. The Baron could for example measure his own speed relative to the cannonballs he encounters. This amounts to assuming that the projectiles coming towards him are at rest. Alternatively, he could postulate that all cannonballs travel at the same speed. Based on this he can then determine his speed by observing how long it takes to get from his own cannons to those of the enemy. Another possible choice is to assume that he himself is at rest and both the enemy cannonballs as well as the cannons themselves move at a relative speed to him. And then there are infinitely more other choices where both cannons and cannonballs all are in motion. All these possibilities correspond to choices of different inertial frames. In our sporting event analogy they correspond to mounting the camera to the Baron himself, the cannon, or the cannonballs shot from the enemy. All these perspectives are valid perspectives, and none is privileged over the other. From each camera perspective, from each inertial frame, we can reconstruct every other perspective by performing a coordinate transformation, by mentally changing the inertial frame. There is no way to designate one or the other choice as the "true" choice.

What this means is that in the ideal Newtonian world, there are no preferred inertial frames. All possible choices of inertial frames are equal. Some may be more useful than others for the purpose of calculations, but there is no possible rational argument to say that one particular special inertial frame is more fundamental than others. Most of all, there is no possible measurement the Baron can perform to determine his absolute speed within the Newtonian world. His speed always needs to be defined relative to some observer. Absolute speed is therefore a meaningless concept in a Newtonian world. This implies that there is no difference between a body at rest and one that travels at a constant speed in absolute terms. In a Newtonian world one can always choose a frame of reference in which an object is at rest. Simply mount a camera on it. Seen from this perspective the two versions of Newton's first law above simply coincide. One can always transform a body at rest into a body at constant speed, simply by changing the choice of inertial frame. For the tale of Baron von Münchhausen this means that he may have just been sitting on a cannonball at rest, but chose his inertial frame in a clever way to make his stories more appealing to the audience.

Relativity

Münchhausen had no way to decide whether he was on a cannon ball at rest or whether he was travelling at great speed. The same would be true for ourselves, at least if we lived in an ideal Newtonian world (which incidentally would not be very amenable to the emergence of life). If you were somewhere freely floating in black space, you would not be able to measure your speed. Now, here it gets interesting. If you cannot decide on your speed, then the same is true for nature itself. Instead of yourself, imagine any physical object in space, freely floating. Just as Münchhausen could not decide his own speed, so is it impossible for any other object or process to "decide" on its speed. In other words, there is no way to derive an absolute speed from observing physical processes. If Münchhausen was not riding on a cannon ball, but instead rode on a floating physics laboratory freely floating in space, no matter what experiment he conducted in his laboratory, he would not be able to deduce his speed from that.

This has an interesting consequence: The principle of relativity. It states that the laws of nature are the same in every inertial frame. Laws of nature have the same mathematical form independent of the inertial frame in which they are formulated. It has to be that way. If there was a change in the physical laws depending on the absolute speed then this change could be used to determine this absolute speed. As soon as one buys into the idea of a featureless space in which the motion of objects can only be determined relative to one another, then it is difficult to escape the conclusion that the principle of relativity is correct.

In practice this means that physics is the same no matter what speed we are travelling at, as long as we travel at a constant speed. The laws of electrodynamics are the same whether we travel fast or stand still, as long as we are not accelerated. In fact, any experiment described in the physics textbooks is true, at high speeds and low speeds. It does not matter.

The principle of relativity is ultimately based on a negative statement: "The absolute speed cannot be measured in a Newtonian world." There are two possible ways to interpret this statement. Firstly, one could assume that there is, in some fundamental sense, an absolute speed, but it is impossible to determine it experimentally. Alternatively, it could be that in some fundamental sense absolute speeds do not exists. So, our inability to measure is not simply a limitation of our methods but nature is such that there truly does not exist an absolute speed in the world.

The difference between these two possibilities is quite subtle. To understand it imagine a huge prison consisting of rows and rows of cells, with no windows for light from the outside, perfectly sound insulated and impermeable to any signal or information from the free world. For the inmates of this place it is impossible to decide whether or not their prison is on a remote island, underground, in the middle of a big city or set in a beautiful countryside. They cannot hear the birds' song in the spring, the cars that drive around the high walls, they cannot know about the seasons, whether it is deep winter or hot summer, whether it is raining, or the sun shines, whether the day has broken or whether they are in permanent darkness. Whatever measurement they make within the prison, they would not be able to deduce what is outside its walls. Yet, even though no information percolates through to the prisoners to them it is still a meaningful question to ask about the surroundings of the prison. They may well want to know where they are and what it looks like outside. These are meaningful questions, only that they are prevented from ever finding out.

It could be that the question about a preferred inertial frame is like this and nature just refuses to reveal a response even though it has one. On the other hand, it could be that there truly is no answer to the question, and nature has nothing to reveal. In the prison analogy this would correspond to the case where the world itself was the prison and everything there is are just blocks of cells, one stacked above the other, with no end or boundary. In this case, it would not be meaningful to ask about the outside of the prison, because it does not exist. From the point of view of the prisoners, one would think, these two possibilities cannot be distinguished.

Debating the existence or non-existence of an unmeasurable quantity seems futile at first, like debating whether or not the moon exists when nobody is looking at it. In physics things are sometimes a bit more subtle and it can be the case that undetectable phenomena make themselves felt indirectly. The idea is as follows: Assuming A is unobservable, one could ask whether A exists, but is unobservable, or whether A is unobservable because it does not exist. It could now be that A, even though it is unobservable, would by necessity imply the existence of a second observable phenomenon P. On the other hand, non-existence of A could imply phenomenon P* If this was the case, then one could experimentally decide whether or not A exists, even though the phenomenon itself is unobservable (and neither P nor P* allow one to make quantitative inferences about A).

An important area in physics where this idea of indirect inference of unobservable phenomena has become important is quantum mechanics. Specifically, in the famous Einstein-Podolsky-Rosen experiment a pair of particles is sent in two opposite directions. The setup is such that these particles can have complementary properties, say either "up" or "down." The rules are that, if one particle is "up," then the other one must be "down." Assume now that each of this pair of particles is sent in opposite directions and somebody then decides at some point to measure the particle that went, say, to the left. She measures this particle and notices that it is, say, "up." Without performing any measurement, she can then immediately infer that the other one must be "down." The question is now the following: Has the particle been "up" all the time, even before it was measured, or has it only "decided" that it is up at the moment of the measurement itself. This question seems unanswerable, because we cannot know what state the particles were before we measure them. Hence, it seems to be quite meaningless. However, physicists could show that it is not a meaningless question after all and there are some subtle observable differences between these two differences that manifest themselves in the measurements. It seems now quite certain that the particles indeed only determine their state at the moment of the measurement. Before the measurement process has started it is undecided which of the particles is up and which one is down.

With this in mind, let us now return to the principle of relativity and its consequences. The question we could now ask is whether measuring an absolute speed is only impossible, or whether the absolute speed is fundamentally undefined. As it will turn out below, it seems that the latter is true. It is not only the case that von Münchhausen is not able to determine whether he is at rest or indeed riding his cannon ball at great speed, but this is fundamentally undefined. Or to use our prison analogy: It is not just that the prisoners do not have any information about the outside world, but this outside world does not exist.

At least implicitly the principle had been known since Galilei. Nevertheless, it was not before the 20th century that its profound implications had been realised. Einstein's theory of relativity is the result of a careful analysis of the implications of the principle of relativity and how it relates to the measurement of speeds in inertial frames. The question Einstein asked was the following: How do we know that two events happen at the same time. Intuitively, this question seems nearly trivial at first—but it is not.

For physicists any type of statement about nature will ultimately have to be formulated in operational terms, that is in terms of concrete protocols for measurements. This also applies to the concept of simultaneity. In order to be able to confidently state that two events happen at the same time, we need to describe a concrete way to experimentally determine the simultaneity of these events. To be able to do this, we first need to have an understanding of how time can be measured. It will turn out below that a careful analysis of the measurement process will lead to some unexpected complications.

Time

We normally perceive the flow of time as independent of the circumstances we are in. The rhythm of day and night (at least in most parts of the world), the change of the seasons, the cycle of birth and death are all obvious manifestations of this thing called time. We do not need clocks to have a feeling for it, but by using clocks we can quantify exactly how much time has passed from one moment to the next.

The subjective perception of time depends not only on the amount of time that has actually passed, but also on how we feel about the events that take place during a given time interval. An hour in the waiting room of a dentist tends to feel much longer than an hour at a fun fair. Despite this psychological effect that distorts the perception of time, in our daily lives we have a natural expectation that a given interval of time is in some sense objectively defined, and equally long everywhere, for everyone and under any conditions, even if it does not always feel like this. If a recent mother states that she spent 14 hours agonising in labour pain, then independently of how long this may have felt to her, we have an objective idea of how much time actually passed from the onset of labour to the first scream of the newborn. A transcontinental trip, say from London to Taipei, that also lasts 14 hours would have a very different psychological impact on the traveller than the birth experience has on the mother. Despite the obvious differences in our perception of time, we take it for granted that there is a reality to the existence of time. Time passes and 5 minutes of objective time are the same everywhere for everybody. So one would think at least.

As it will turn out, this is not true. A careful analysis of the operational concept of time, which forms the core of Einstein's theory of special relativity, will show that our intuitive beliefs and assumptions about time concepts are wrong, in the sense that Einstein's theory predicts phenomena that are at odds with our inherent expectations. One of the main results of the special relativity is that the length of a time interval depends on the inertial frame from which it is measured. Two observers judging from different frames of references will judge the length of one and the same event differently. This difference in time has nothing to do with the subjective experience of time, which differs from person to person, and from situation to situation. The two observers will both have different objective measurements of time. In what follows we will always refer to objective time, as measured by ideal, unfailing clocks and completely ignore subjective perceptions of time. This does not mean that subjective time is not a reality, but we choose to ignore it.

To understand the basic ideas of Einstein's special relativity we now take a step back and ask a specific question:

Assuming that the woman was in labour for 14 hours and assuming that the transcontinental traveller was travelling for 14 hours: "Are 14 hours of labour the same amount of time as the 14 hours of travel?"

At a first glance, this seems to be a silly question. It is an obvious truth that 14 hours are always 14 hours. Evidently true as it may seem, we have indicated above that the answer to this "silly question" is not at all straightforward. In a strict sense, 14 hours will always be 14 hours, yet there is a way in which we can reformulate the question slightly, so that what seems so obviously true becomes false. The 14 hours of the traveller and the labouring woman may indeed, in a very objective sense, be different amounts of time—and we mean here different amounts of *objective* time, not subjective time. How can that be? To develop an understanding of this it is first necessary to look at the concept of time in some more detail.

Measuring time

In order to approach the "silly question" above, let us start with a simpler question first: How do we know how much time has passed since an event. How does the labouring woman know that she has been in labour for 14 hours, rather than 1 or 12 or 20 hours? The simple and only answer to this is that she used some sort of clock or timing device. To make it concrete, we assume she had a stop watch with a red button. At the first sign of labour she presses the button to start the clock. The moment the baby takes his first breath she presses the button again to stop the timer. We further assume that the timer stopped precisely at 14 hours, 0 minutes and 0 seconds.

Saying that she used a clock does not solve the measuring problem because it only shifts the problem of determining the length of a period of time to the problem of building a clock, which arguably is the same problem. So, the next question to ask is how to build a clock? How can we be sure that any particular timing device actually measures time correctly?

To understand this it is instructive to consider a simpler problem first: Measuring length. The principle there is very simple. One could take a stick made of some durable material, cut it to a convenient size and define the length of the stick as the unit of measurement, for example 1 meter. This is precisely the way the so-called ur-meter has been used, a stick made of metal with two marks on it that defined the standard for a meter. Until 1960 this ur-meter was kept at a research institute in Paris and served as the standard unit of length. In practice, one would have to worry about variations of the length of this ur-meter as the temperature changes, but we ignore this and similar technicalities here. In principle, once one has this ur-meter one can measure any object simply by checking how many copies of this ur-meter are required to fit the object to be measured. So, measurement of length is nothing but relating extensions of bodies to some known standard that has been arbitrarily fixed at some point. Measuring length is conceptually easy, although in practice somewhat more complicated than indicated here.

We could now try a similar procedure for the measurement of time and define a reference interval of time, an "ur-second" or "ur-minute" which can then be used as a reference for any other interval of time. There are many possible ways to define such a measure. One could for example use the time from sunset on one day to sunset on the next day as a unit of time. For some purposes this will be useful, but mostly we would expect it to be imprecise because of natural variations of the daylight hours over the year.

Another simple possibility to measure an interval of time is an hourglass. An hourglass consists of two transparent cones that are connected at their tip by a small opening. If the cone on top is filled with sand or salt or grains of any type small enough to fit through the small opening, then over time, all the sand will move from the upper cone to the lower one. As a technology, the hourglass is very simple, but still finds applications in kitchens and saunas and similar areas. One could use such an hourglass to define an "ur-minute" as the time required for all the sand to fall from one cone to the other. To measure the length of say the labour of the mother to be for example one would then have to count how often it is necessary to turn the hourglass before the child is born.

The hourglass is only useful as a measure if the time to empty one cone is constant, that is each time it is inverted it takes the same amount of time to empty it again[1]. At this point we encounter a problem. How can we be certain that the hourglass period is always the same, that the time corresponding to one hourglass cycle does not vary? Based on our understanding of the mechanism of the device it is certainly reasonable to assume that one period of the hourglass is the same as the next period. However, we cannot be certain of the accuracy of the device unless we have another timer.

Therefore we need another timer to check the hourglass. One

possible alternative is to use the pulse beat as a putatively regular process. The experimental protocol is simple. All we need to do is to count the number of pulse-beats per hour-glass period once and then once more and compare the outcomes. In practice, this experiment should be done not only twice but several times. After a series of comparisons between hourglass periods and pulse beats there are then several possible outcomes. Either the hourglass period always corresponds to the same number of pulse beats—or it does not.

When there are deviations between successive trials, these deviations could be large or small. Precisely what counts as large or small is a bit fuzzy. In practice we would know that the deviations are large if, for example, one trial yields 140 beats per ur-minute and the next one 247, a third one only 14 and so on. Differences of this sort would immediately be worrisome and we would be forced to conclude that either the hourglass or the pulse or both are unsuitable as timers, simply because they cannot be brought into agreement with one another. In this case, one would have to find other timers.

Let us now suppose that the deviations from one trial to the next are very small. In this case one could hypothesise that the discrepancy was due to some random small errors rather than an underlying fundamental irregularity of either process. The error could be due to a number of different factors. For example, simple counting errors could be responsible for the deviation between results. Another source of error could be the non-constancy of the pulse-rate. The experimenter may have gotten up between experiments and made herself a cup of tea which altered her pulse rate. Another possibility is that the pulse rate remained constant but the hourglass period varied across repeat trials. We cannot know for sure which of these possibilities applied. Yet, the experiment showed, if nothing else, that at least one of the two measurement devices is intrinsically limited in its accuracy. Hence the error between repeat trials.

In any real life measurement situation random noise will invariably affect the result, and there is no way to get around this. Leaving this aside for the moment, let us assume that the series of comparisons between the hourglass and the pulse led to precisely the same pulse per ur-minute count in each and every single trial. In this case, we would certainly feel entitled to assume that the hourglass and the pulse-beat are reliable measures of time. Unfortunately, the situation is not as straightforward, and loopholes remain. It could be that the hourglass and the pulse agree with one another because they are affected by the same distortions that prevent them from measuring the "correct" time. To some extent one can guard against this by validating the hourglass and the pulse with yet another and another candidate timer. Each time a new such timer is evaluated, it could either be in good agreement with all the other—already established—timers, or not. In the latter case, one will tend to discard the candidate system as a useful timer, but accept it otherwise.

This procedure of adding new timers and cross-evaluating them is not in any sense a logically valid procedure. The common bias argument can never be avoided, no matter how many timers there are on the list. It may also be that some of the discarded candidate-timers may actually be the true measures of time, whereas all the other apparently good clocks are, somehow, fundamentally flawed. However, as the list of timers that are in agreement with one another grows, and as these timers prove themselves practically useful, there comes the point when the loopholes cease to matter. Time is not a tangible quantity that can be directly detected like the number of grains in a bag of sugar. We cannot know whether or not our timers capture time correctly, because we have no access to time by means other than our timers. An operational notion of time is always based on recording correlations between physical processes. We cannot know that some events are regular unless we compare them to some other putatively regular events, and observe correlations between these processes. Modern wristwatches are based on this principle just as hour-glasses and super-accurate atomic clocks. Mechanical watches exploit regular movements produced by clever arrangements of springs and cog-wheels, and quartz watches use the oscillations of a piece of quartz crystal. Based on careful observations we feel confident to assume that there are clear relationships between the period of a quartz crystal and that of a spring-driven cog-wheel.

We conclude, that measuring time really means measuring physical processes that proceed in a regular fashion. We conjecture that processes are regular based on fixed relationships between them. It is not possible to get independent access to time, in the sense that there are independent means of checking that the supposed regularities are indeed regular in time. This has an interesting consequence: If time was to slow down or speed up, we would not notice it, at least not if all processes in the universe are equally affected by this speeding up and slowing down. Any change of the flow of time would affect our bodily processes, the speed of our thoughts or any other event in the same way as it would affect the frequency of the quartz crystal in our watch and everything else. When we attempt to reason about such uniform changes of time, we even start to reach the limits of language in the sense that it is impossible to speak about the flow of time and its changes without using concepts that themselves presuppose a regular ticking of a clock (such as "frequency" and "slower" in the previous sentence).

For the physicist this means that it is meaningless to talk about time, other than by referring to clocks in some way, which in turn means to count some regular physical process that has previously been shown to be in reasonable good agreement with other putatively regular processes. The upshot of this is that time is only a reality insofar as it can be measured. Time and the measurement are inextricable and there is no meaningful way in which the quantity "time" can be separated from the act of measurement of time.

Time dilation

With these preliminaries in place, we are now in a position to return to the "silly" question we asked above: "Are the 14 hours of labour the same as the 14 hours of transcontinental travel?" In order to address this question from a physics point of view, we have to formulate it in terms of measurement processes. There are several ways to do this. One possible experimental protocol to follow is this: As soon as labour sets in the woman presses her timer and at precisely the same time the traveller starts his. At the first scream of the baby, the mother then turns her timer off and exactly at the same time the traveller stops his watch. We then ask whether both timers show exactly 14 hours 0 minutes and 0 seconds, or whether the two clocks show different measures for what seems to be the same interval of time.

One problem both the labouring mother and the traveller will face is the inaccuracies of their measurement devices. Every measurement is afflicted by inaccuracies. Since measurement errors are random they will most likely affect the two measurements differently, so there will be discrepancies between the time of the labouring woman and the time measurement of the traveller. These errors are just a normal part of the life of experimental scientists and anybody else who measures things in the real world. They do not point to any fundamental effect. Theoretically, they are therefore not very interesting, at least not for our purpose here. We will therefore just ignore them. In any case, we are only engaging in *gedankenexperiments* where we can abstract away from all the nuisances of real life.

There are a few other complications that we need to address. The most important complication, that will ultimately take us to the heart of the issue, looks at first quite innocuous. A key element of the experimental protocol is that the woman and the traveller switch on their timers *at the same time*. This is only possible if the traveller knows the exact moment of the onset of labour. In our protocol above we have simply assumed that there is a way for the traveller to turn on his timer simultaneously with the woman starting hers, but we have not specified how he can know that the woman has switched on her watch. This needs to be fleshed out in more detail if we are to describe a workable measurement protocol.

One possibility is that the traveller is with the woman right until she starts her timer; at this point he will then also start his and leave for Taipei. This is a possible scenario, but not quite what we want. The problem is that "leaving for Taipei" implies an accelerated motion. Concretely this means that the traveller will not always be at the centre of an inertial frame. This is an inconvenient scenario because it violates our simplifying assumption that both our protagonists live in a Newtonian world and that they move at a constant speed to one another[2]. We will therefore choose a different scenario that is a bit contrived but will prove to be simpler ultimately. We assume that the traveller and the woman are at rest in two different inertial frames that are in constant motion relative to one another. Now we have shoehorned our supposedly "real world" example into a scenario fitting the assumptions of the Newtonian world. Theoretically this is very convenient and it highlights one aspect of the scenario that will prove to be the origin of the theory of relativity: The woman and the traveller need to remotely *synchronise* their clocks. In the context of our scenario "synchronising the clock" means simply that there is some way to ensure that the timer of the woman and the traveller are started at the same time. The most efficient way to do this is to send a radio signal from one device to trigger the other one. What makes radio signals so convenient is that they travel at the speed of light. In fact, radio signals and light are fundamentally the same phenomenon in that both are electro-magnetic waves that can travel through empty space. It is commonly thought in physics that the speed of light is the maximum speed with which anything can travel, including information. So, it is not possible to transmit a signal from A to B faster than the speed of light. This is a fundamental assumption underlying Einstein's theory.

Let us make our scenario a bit more concrete. We can imagine the timer of the traveller to be a handheld device that can receive signals directly from the clock of the labouring woman. The woman has a similar device. Once the labour starts, she presses a button which then causes her device to send a signal to the traveler. When he receives this "Go" signal his timer will switch on and continue until it receives a "Stop" signal once the baby is born.

Now, with this setup we immediately encounter one problem. The synchronisation signal travelling at the speed of light is fast, but not instantaneous. It will take a certain period of time for the signal to reach the traveller. This delay needs to be taken into account which is not a problem as long as the position of the woman and of the traveller are known. We will return to this point in a moment. For now let us take a step back again and reformulate our "silly question" to make it more precise:

Assume a woman who is in labour for 14 hours, and a traveller who moves at a constant speed relative to the woman. Assume further that the woman starts a timer at the onset of her labour and stops it when the baby is born and that she sends a "Go" and "Stop" signal at these two events to be received by the traveller. Will the clock of the traveller, after receiving the off signal and after correcting for delays due to the travel time of the signal, show exactly the same time as the clock of the labouring woman, namely 14 hours exactly?

To pre-empt the answer to the question: No, it will not. The 14 hours of the labouring woman will not be the same amount of time as the 14 hours of the traveller. Depending on the speed of the traveller, the difference may not be big, although for high velocities it could be. This is one of the key-results of Albert Einstein's theory of relativity that showed that moving clocks tick differently—quite literally.

This fundamental result can be derived by considering how precisely time periods are measured and how they can be compared experimentally. A careful examination of what needs to be done to ensure that the clock of the woman and the traveller are started at the same time, will lead us to examine what precisely it means that two things happen at the same time and how this can be ensured experimentally.

In the example of the woman and the traveller, there are three things that need to be determined. Firstly, the woman needs to measure the length of the time interval, from the onset of labour to the birth of the baby. The traveller needs to measure the time between the two radio signals he receives from the woman, and thirdly the traveller needs to correct for the time delay that is caused by the time it takes for the radio signal to reach him. The first and the second point seem unproblematic. Measuring time intervals is not entirely trivial, but we can consider the problem solved and assume that both of our protagonists have infinitely accurate timers at hand.

We will therefore focus on the process of synchronisation and ask the following question. When the woman turns on her timer and emits the synchronisation signal to the traveller, how long will it take for the traveller to receive the radio signal? This is quite straightforward to work out. We will not write down the formula here, but it is easy to see how it can be done. The radio signal travels at the speed of light, which is about 671 million miles per hour. If the distance of the traveller from the woman at the time she presses the button is known then it can be quite easily calculated how long it takes for the signal to travel. There is one complication, namely that the traveller moves relative to the woman. This adds an extra delay time to the synchronisation time, because the radio signal needs to catch up with the traveller. Transcontinental aircrafts have a maximum speed of about 640 mph which is roughly a million times slower than the speed of the radio signal. Assuming that the traveller is stationary for the purpose of the calculation is probably a very good approximation in this case, although strictly speaking there is some extra delay due to this movement, however small it may be at realistic travelling speeds.

At speeds we normally encounter we are best advised to ignore this delay completely. But let us now take advantage of the freedom of gedankenexperiments and probe more dramatic scenarios. For the sake of the argument assume that the traveller moves at speeds relative to the woman that are comparable to the speed of light. Currently it is technically impossible to construct aircrafts that achieve velocities anywhere near the speed of light, although there are no fundamental limitations that prevent future engineers from building spaceships that can approach, but never reach, the speed of light. It is therefore perfectly legitimate for us to assume that the traveller uses one of these ultra-fast futuristic modes of transports. Once we grant ourselves this freedom, then this has a very tangible consequence for our synchronisation problem. When we calculate the correction due to the travelling time of the radio-signal, then we can no longer ignore the speed of the traveller himself. There will be an extra delay due to the signal catching up with the movement of the traveller.

This can be easily illustrated using every-day examples. Consider a car chase: Two cars, one in front (the traveller) and one behind (the radio signal) trying to catch up with it. Let us assume that the pursuing car is faster than the car it tries to catch. Let us for the moment also assume that the car ahead is only slightly slower. We can now ask how long it takes for the pursuer to catch up with the pursued. Clearly, this will depend on the distance and speed difference between the cars. To make it concrete, assume that the pursuing car has a speed of 100 mph, and the chased car has a speed of 40 mph. The chasing car will then approach the chased car with a speed of 60 mph. If the distance at the beginning of the chase was exactly 1 mile, then it will take the pursuer exactly one minute to catch up. If however, the chased car travels at 70 mph, the speed difference is only 30 mph and the pursuit will last for 2 minutes. Finally, if the cars race at precisely the same speed, then the chasing car will never catch up with its target.

This very same intuition can be used to understand how the radio signal catches up with the traveller. The motion of the aircraft relative to the woman causes a delay. The higher the speed of the traveler relative to the woman, the longer the synchronisation will take.

So far we have looked at the situation from the perspective of the labouring woman. We have tried to understand how long it takes for the synchronisation signal to reach the traveller. We can now change perspective, change so to speak the inertial frame and ask what the traveller sees. Based on our everyday experience, we would assume that he would report the same delays as the labouring woman. Certainly in the car chase above it should not matter whether we ask the chased driver or the pursuer. They would both report the same speed difference and if asked, they would agree about the time the catch-up took.

Here is the crux of the argument. This intuition only holds at low speeds, that is speeds that are low compared to the speed of light. The crucial point is the principle of relativity. According to this principle the laws of nature are the same no matter what inertial frame of reference one uses. Now, remember that we assumed that the traveller moves at a constant speed. So, he can be thought of as the centre of an inertial frame, just as the labouring woman can be thought of as the centre of a different inertial frame. For both the woman and the man, the laws of physics must be the same. In particular, this means that the speed of light must be the same for both of them. This means that both the labouring woman and the traveller would both always observe that light and radio signals travel at the speed of light.

A moment of thought will now show that this implies a

discrepancy in their judgements about the required synchronisation time. The woman would have to conclude that the light signal needs to catch up with the traveller. Just as in the car chase, if the traveller is only a bit slower than light, the catching up time can be very high. From her perspective the signal approaches the traveller with a speed corresponding to the *difference* between the speed of light and the velocity of the traveller. The traveller on the other hand, observes no such extra delay. He sees the signal approach him at the speed of light, and *not* the difference between his own speed relative to the woman and the speed of light. Consequently, from his perspective, the synchronisation signal arrives much faster than from the perspective of the labouring woman. Two observers come to possibly very different conclusions about the duration of one and the same event. Ouch!

Just to see how this clashes with our every-day experiences, let us frame this within the car chasing model. The situation there would be something like this: The chasing car drives at 100 mph and the chased car at 50 mph. An onlooker or a camera fixed to the track (corresponding to the woman) would observe the relative speed between the two cars to be 50 mph. However, from the perspective of the pursued car (the counterpart of the traveller) the speed of approach would always be 100 mph, no matter what its own speed is. At least this would be the case in the world where the law of nature states that pursuing cars always approach at 100mph.

The principle of relativity and the constancy of the speed of light leads us to a situation where two observers of one and the same event come to entirely different conclusions about the time it takes for this event to complete. This is a very unusual proposition and seemingly contradictory. That is, contradictory as long as we assume that time flows equally, for both the woman and the traveller. This assumption seems to be a reasonable one. Given our daily experience we have little reason to assume that it is wrong. Yet, if we are prepared to give it up, some of the conundrum vanishes.

If we allow for the possibility that time passes much slower for traveller than for the woman then the apparent contradiction disappears. The traveller perceives the synchronisation signal as travelling much faster than the woman because in his inertial frame time passes slower. In his frame of reference clocks and all physical processes are slowed down relative to the inertial frame of the woman. Every second in her system corresponds to a fraction of a second of his. Whenever he needs to turn an hourglass timer, she would have turned them before; maybe even many times. It sounds odd, but it makes the physics consistent again.

Physicists call this effect *time dilation*. We will not worry about the precise formula from special relativity that determines time dilation, but we can consider two cases: When the traveller is at rest relative to the labouring woman, then time passes at exactly the same pace for both. In the extreme case that he travels at the speed of light, time would stop in his inertial frame. This corresponds to the case where the radio signal emitted from the woman would never catch up with the traveller.

Time dilation can only be observed when two events in different inertial frames are compared. As long as we remain in a particular inertial frame, we do not notice anything special. In fact, we cannot even determine in any reasonable way whether or not we are fast or slow because there is no way to determine absolute speeds of inertial frames. The traveller himself would not notice the slowing down of time. Relative to the system of the labouring woman everything would have slowed down. The world, his experiences, his thought processes, the ticking of the clock and any other aspect of his existence. This includes any regular physical process that he may exploit for time-keeping. So, when everything slows down so will his clocks. This slowing down cannot be experienced. The traveller, just like the rest of us, can only know about time by comparing it to some regular phenomena, or by measuring them with watches.

He would only notice the effect of time dilation if he came back to the labouring woman, and found that the child she has born when he left a few days ago is now a grown woman and may be in labour herself. What felt like a few days for him, indeed what was only a few days for him may have been half a life-time for the woman. The silly question was not that silly after all, it was only ill-posed.

This is the conclusion of Einstein's special theory of relativity. Einstein himself reached it mostly based on theoretical considerations. The question now is whether the *gedankenexperiment* and its conclusions are anchored in reality. Is there indeed a time dilation?

Experimental Evidence

Using sophisticated atomic clocks and all the fancy equipment that modern science and technology has produced, physicists can nowadays indeed measure time dilation. However, at the time of Einstein this was well beyond the technological possible. The effect is very small unless one travels very close to the speed of light, hence very difficult to measure. However, there was some other experimental support for Einstein's theory before time dilation could be measured directly.

This effect that clocks tick differently in different inertial frames rests crucially on the fact the assumption that the laws of nature look the same in all inertial frames, or equivalently, that it is not possible to determine an absolute speed in a Newtonian world. Contemporaries of Einstein were concerned with this question even before he came up with his theory of relativity. Einstein himself refers to the experiment by Fizeau who measured the speed of light through a pipe of flowing water. He hoped to detect some difference when the light is sent with and against the flow of water. Today another experiment is much better known. The Michelson-Morley experiment (or MME) originally performed by the American scientists Albert Michelson and Edward Morley at the end of the 19th century is usually cited as the key experiment to support the theory of relativity. The original motivation for the experiment was not related to relativity at all. Instead Michelson and Morley set out to detect the existence of the *aether*, a hypothetical and somewhat mysterious lumniferous form of matter that pervades the universe. For a long time physicists believed that the wave-like nature of light required some medium that could propagate the waves, just as sound waves are carried by air. Aether had never been experimentally verified and it was also postulated to have some rather strange properties. Aether had to be massless, incompressible and also of zero viscosity. It seems rather strange today. Nonetheless, contemporaries of Michelson and Morley mostly believed in its existence. The MME was an attempt to finally prove its existence experimentally, although by indirect means. The basic idea of the experiment is fairly simple. If aether existed then the rotation of the earth should create a relative motion to it. Since aether was assumed to be the medium that propagates light-waves, this relative motion of the earth should manifest itself in different speeds of light in different directions. So, the speed of light along the earth's rotation should be different from the speed across the rotation.

A suitable analogy are water waves. If one sits on a jetty and measures the speed with which the water-waves below roll to the beach, one will get a different result than if one observes the very same waves from a motor-boat that goes out to sea, and again a different result if one measures them from a boat coming to shore. Since the boat moves relative to the waves, from the perspective of the boat the waves seem to travel at a different speed than they would from the jetty. This is again all about frames of reference. Another way to think about this is to imagine a river, with a small but steady flow. If suddenly somewhere a fish comes up to the surface, or one throws a small stone into the river, then a circular pattern of waves is created. This pattern of waves travels with the flow of the river, but at the same time expands in all directions. Seen from the sides of the river the waves travel much faster along the direction of the flow, than against it.

Exactly the same idea underlies the MME. The rotation of the earth creates a relative motion against the hypothetical aether which should be detectable if the aether really existed. This relative motion is different along the direction of the rotation (east-west) than it would be perpendicular (north-east) to it. This difference should impact the speed of light, just as the flow of the river would manifest itself through different wave propagation speeds along and against the flow.

The MME was never designed as a crucial experiment to test Einstein's relativity, indeed it preceded it. Michelson received the Nobel prize for his contributions to optics in 1907, two years after Einstein's first publication on the special theory of relativity. However, the original MME was published in 1887, well before special relativity was known. Despite preceding the theory, the MME is highly relevant. It tests directly the constancy of the speed of light and as such the foundations on which the theory of special relativity is built. Aether theory introduces a preferred inertial frame. If aether existed and is measurable then it would be possible to measure absolute speeds relative to the aether. In this case, the whole reasoning behind Einstein's theory of relativity would come down.

The expected difference in speed along the east-west and the north-south axis is extremely small compared to the absolute speed of light. Direct measurements would require levels of accuracy that are nearly impossible to achieve. Sensing the effects of earth's rotation on the speed of light would be a bit like trying to detect a single mouse in a cruise ship by determining its weight. The largest cruise-ships have a displacement of about 100000 tons. A mouse has a weight of about 100 grams at most.

So, even if the exact weight of the cruise-ship is known, it will be very difficult to measure it accurately within a few grams. Similarly, extremely accurate measurements of the speed of light would be required to reduce the experimental error sufficiently to detect the relative motion of the Earth against the aether. This is difficult today and would have been even harder at the time of Michelson and Morley. They found an ingenious way around this. Knowing that they could not measure the absolute speed of light accurately enough, they realised that they could measure the difference of the speed of two beams to much higher levels of precision. In the analogy of the cruise ships this is easy to understand: Instead of measuring the weight of a cruise ship directly, they took another identical cruise-ship that is known to be rodent-free and then compared both ships by putting them on a finely gauged balance scale. If they balance perfectly then there is no mouse on board, otherwise there is. This method requires much lower accuracy.

As an aside, it is interesting to see that they followed a very similar strategy to Galilei. Just like him, Michelson and Morley were faced with an experimental design challenge that was just about feasible given the technological means of the day. In both cases it was key to control the experimental error in order to get meaningful results from the experiment which was best achieved by indirect methods of measurements. Galilei had to prove his square law by considering a related system and arrive at his conclusions from there. Michelson and Moreley pursued a similar strategy.

Concretely, they exploited the wave property of light. Light is an electro-magnetic wave and can be thought of as a travelling pattern of peaks and troughs, very much like water waves. Interference is a phenomenon that may occur when two (or more) such travelling patterns of peaks and troughs are brought together. If every peak of one wave coincides with a peak of the other wave then together these peaks will form even bigger peaks. On the other hand, if each peak of one wave coincides with a trough of the other, then the two waves may cancel one another out, resulting in no wave at all. In water waves such interference events can lead to extremely high waves on oceans, but could, in the case of negative interference, also produce a seemingly calm sea. Just as in water, interference between light beams can be demonstrated experimentally. These interference patterns manifest themselves in characteristic stripes of light and dark circles. The dark circles correspond to negative interference and the light circles to areas where the troughs fall on troughs.

These days the experimental setup of Michelson-Moreley is a standard lab-exercise performed by physics undergraduates in their first year. Using common laser-pointers it is now even possible to build a simple Michelson-Moreley interferometer at home on the desk; instructions are available on the internet. With modern technology, particularly lasers, producing interference is not particularly difficult any longer, although not entirely trivial either. The basic idea is to split up a beam of light into two beams, send the two beams along perpendicular directions, and re-unite them. The simplest way to split a beam is to use a semi-transparent mirror that lets half of the light pass through while reflecting the other half, plus 2 more fully reflecting mirrors to control the two paths of the light.

Let us assume that the two paths along which the beams travel are exactly of the same length. If we re-unite the two beams again then we can be certain that each trough falls exactly on a trough and each peak falls on a peak. We have re-created the original beam. There is one caveat. In order for this to be true we also need to assume that the speed of light was the same in both directions. If that second assumption does not hold, then the peaks and troughs of two beams will not necessarily be perfectly aligned, even if the path-length was exactly the same for both. Experimentally it is very difficult to be certain about the first assumption, namely that the paths are equally long. In most cases it would be very difficult to guarantee this. Luckily, the assumption is not needed for the MME. If light is sent through a Michelson-Moreley interferometer, then one will observe a characteristic pattern of light and dark patches which is a result of the different path-lengths of the two split beams and—so Michelson and Morley hypothesised—of the difference in the speed of light along the two paths.

Given an interference pattern from a Michelson-Moreley experiment one cannot draw any conclusions about the difference in the speed of light between the two paths. The pattern will depend both on any differences in speed and difference in path-lengths. It is impossible to disentangle the contributions of each. In a sense, there are two unknowns. Now comes the crucial insight by Michelson and Morley. If they turned the interferometer, then this does not disturb the internal geometry of the setup. Hence, the path-length of the individual beams remains the same. So, one of the two unknowns remains unchanged. The putative difference in the speed of light between the two beams, on the other hand, may depend on the orientation. As the interferometer is turned, the angle of the light paths with respect to the hypothetical aether wind is changed. This should shift the interference pattern in a predictable way. So, as the device is turned, any change of the interference pattern would be due to differences in the speed of light.

What sounds easy in theory is difficult in reality. Turning a highly sensitive interferometer without introducing changes to its geometry, subtle changes to the orientation of the mirrors or the path-lengths requires very careful design and much patience to get the setup working. In the original MME the interferometer was mounted on a turntable suspended in a mercury bath which would dampen out any vibrations originating from the environment.

We will not further describe their experiments. Suffice to say that within the experimental error Michelson and Morley could not prove a difference of the speed of light in the different directions. Effectively, this disproved the aether theory. Consequently, in the aftermath of the Michelson-Moreley experiment the concept was abandoned in physics. However, experimentally they have not shown that the speed of light is the same in every direction on earth; they only demonstrated a lower boundary on the difference. Even though the aether is no longer a contender, there could still be a small dependence of light depending on the speed of the inertial frame, which would be of theoretical interest. Following the original Michelson-Moreley experiment there have been many improvements to the experimental setup reducing the experimental error by many orders of magnitude. All of these modern day Michelson-Moreley experiments always only lower the boundary on the difference down to experimental error. In all those experiments it is therefore crucial to control this error, to screen the experimental setup more and more from environmental conditions: to build better laboratory systems.

Müller-Peters experiment

A modern version of the MME was performed by the German physicists Holger Müller and Achim Peters in 2001. Unlike the original MME Müller and Peters did not use an interferometer and interference patters. Instead they used so-called "optical resonators." In their simplest form these optical resonators are two mirrors facing one another. If a beam of light is coupled into a resonator it will be sent back and forth between these two mirrors. In theory, once a beam has entered the resonator it will be reflected back and forth indefinitely. In practice this is not the case because the mirrors are not perfectly reflective. Some light will be let through, some light is absorbed. This means that any beam will only travel back and forth a finite number of times before it disappears from the system. Müller and Peters used highly reflective mirrors that allowed the beams to go back and forth between the mirrors for several hundred thousands of times.

Within such a resonator there is again the possibility for interference effects. Just as in the interferometer where two beams are split and then re-united to produce interference patterns, in the resonator there are possible interference effects between the two beams that travel into opposite directions between the two mirrors. Depending on the distance between the mirrors the peaks and troughs travelling in opposite directions may coincide. This *resonance* will only happen when the wavelength and the speed of light and the distance between the mirror are in just the right relationship to one another. Change one of them and there will be no resonance. This critical dependence of the resonance effect on these three parameters is the effect that Müller and Peters used to measure the speed of light.

The idea is very simple: Among those three parameters, the speed of light is the unknown quantity they wish to measure. There is nothing they can do about it. The distance between the mirrors is fixed and can be measured quite accurately. Once set, it would not change any more either, at least it should not change substantially. The critical quantity that they could control was the wavelength. Within their setup Müller and Peters could adjust the wavelength at will. This means, that they knew both the wavelength and the distance between the mirrors, which is sufficient to calculate the speed of light.

Their experimental procedure was straightforward: First adjust the wavelength until you get resonance. Then use this information to calculate the speed of light. For their version of the MME Müller and Peters needed to determine the speed of light in two different directions, just as in the original Michelson-Moreley experiment. Following their predecessors they thought that they would be better off measuring differences in the speed of light, rather than comparing two absolute measurements.

This is what they did: In order to realise the MME using optical resonators Müller and Peters had an arrangement of two optical resonators, one of which they aligned along the rotation of the

Earth (east-west direction), the other one they set up at an angle of 90 degree angle to this (north-south). Then they adjusted the wavelength in each of their lasers independently until they observed resonance. Any difference in the speed of light between the two directions should manifest itself in the optical resonance occurring at slightly different wavelengths in the two orientations. In practice Müller and Peters measured the resonance frequency (which is just the inverse of the wavelength) using an electronic control loop that continuously fine-tuned the wavelength of a laser beam pumping the resonator in response to the output of the resonator. For both orientations they then compared the frequencies of the two lasers. They found that there was no difference in the speed of light, at least not within the experimental error. This again confirmed the principle of relativity.

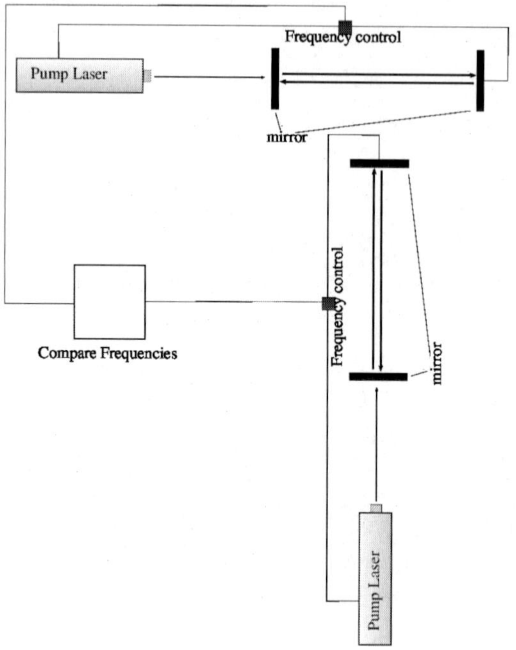

Figure 2.3: The basic experimental setup of the Michelson-Morley experiment by Müller and Peters.

The basic idea of the experiment is simple and contains no ground-breaking new insights. Even before they published their paper reporting the results of their experiments, probably nobody within the experimental physics community would have been particularly impressed or surprised by the idea of performing a Michelson-Moreley experiment using optical resonators. The results of the experiments themselves were not of particular importance either. It is unlikely that Müller and Peters truly believed that their experiment would turn out to overthrow conventional wisdom by showing a violation of the principle of relativity.

The true significance of their work in the context of experimental physics was in improving an experimental technique. The basic principle of the Müller-Peters experiment in fig 2.3 could be realised in an undergraduate physics laboratory course—but, and this is the point, not to the same degree of precision that Müller and Peters achieved. Müller and Peters novel contribution is how they managed their errors and achieved an improvement in the accuracy of the measurements relative to previous attempts. Müller and Peters built better laboratory systems. This is what experimental physics is all about.

In an experiment like the MME which is ultimately about showing the absence of an effect it becomes crucial to control the experimental error. These errors come from many different sources. Each of these sources needs to be found and then meticulously eliminated, or at least suppressed as well as possible. One source of problems are the apparatuses themselves. Measurement devices are always limited in their ability to distinguish between different values. Kitchen scales, for example, have an accuracy of about a gram. If one measures 250g of flour then the actual amount could be 250.7g, 249.8g or 251.4g or any other similar amount of flour. The scale is not designed to distinguish between these values, and they do not need to. Scientific instruments are normally more accurate than that but they all have finite accuracies.

Beside the inherent limitations of the devices themselves, a second major source of experimental error are disturbances and noise from the environment. Anybody who has spent some time at a university department for experimental physics has heard stories about ultra-sensitive measurements that go wrong when somebody passes by outside the lab door. These stories may not be true, and indeed, if they are then they probably point to bad experimental design. In any case, they illustrate that environmental disturbances of all kinds are the enemy of high precision measurements. Changes in temperature, humidity, mobile phone signals, fluctuations in grid voltages and hundreds of other factors that are barely noticeable in real life, can be dominating effects when measurements are to be accurate to a millionth part of a degree. Such external fluctuations of conditions cannot be totally eliminated, but the experimental physicist will spend significant amounts of time to screen her laboratory from such unwanted fluctuations. The skill of the experimentalist is to take hundreds of design choices that together lead to an ultra-sensitive system that is well isolated from the environment.

Since the MME had been performed many times before, the only justification for repeating it is to reduce the experimental error, to reduce the lower bound on the difference between the speed of light in the two directions. So, controlling the experimental error is the be all and end all of Müller and Peter's experiment.

A major contributor to the experimental error in the Müller and Peters experiments are temperature fluctuations in the laboratory. When the temperature changes then materials expand or shrink to some degree. In daily life these changes of the size of objects can make themselves felt, although, the normal fluctuations around room temperature are usually negligible. In the realm of hyper-precise experimental physics, even the tiniest fluctuations are unacceptable. Remember: The whole idea of measuring the speed of light using the resonator is that both the distance between the mirrors and the wavelength are known, so that the speed of light can be determined. When the distance between the mirrors fluctuates, then this introduces an uncertainty in the true value of this dimension which then translates into an uncertainty about the speed of light. Size effects of the materials are not completely avoidable, but can be minimised by screening the system from temperature fluctuations and by choosing suitable materials. Müller and Peters submerged resonators into a bath of liquid helium which is only about 4 degrees above absolute zero temperature (4 degrees Kelvin corresponding to about −269°). This ultra-cold inner core bath of liquid helium was itself contained within a block of copper containing an outer core of liquid nitrogen at about −196° to further shield the experimental system.

This elaborate temperature control was essential to keep the experimental error down but also created its own problems. The experiments by Müller and Peters went on over several months and the coolant liquids had to be replaced about once a day. In the case of the liquid nitrogen this had to be done manually, but was automatic for the helium. The experimental setup was so sensitive that changing the coolants and the subsequent small alteration of the internal geometry within the copper block caused detectable spikes in the output reading. This illustrates just how sensitive the Müller and Peters experiment was. To minimise the disturbances from the periodic refills, they used an automatic beam repositioning system to re-calibrate the system on a continuous basis.

The liquid coolant core isolated the system from temperature changes in the environment, but what it could not do was to prevent the system from warming as a consequence of the internal operation of the system. The laser light that was used to pump the optical resonators carried some energy, some of which was absorbed into the device and converted into heat. This type of heat generation is unavoidable, but can be minimised by using very low powered lasers. Another strategy to minimise the effects of heat expansion was to manufacture the mirrors using materials that show low amounts of shrinkage/expansion with temperature changes. Müller and Peters chose sapphire crystals. Each resonator was made from a single crystal, first cut to suitable shapes, prepared, and then carefully mended back together to form the core of the resonators.

The Müller and Peters experiment is just one of thousands of experiments performed every year in physics departments around the world. It can therefore be hardly considered typical for a physics experiment. In some sense, however, it is representative. It shows how the experimenter battles against the environment, how much time and energy and money she has to use to build the laboratory system, the experimental device that allows her to construct and observe the idealised world of the theoretical physicist.

In many ways the Müller-Peters experiment faces analogous challenges as Galilei's experiments. Despite the superficial differences between these two experiments, there are fundamental similarities that unite the two. For one, in both cases the experiments challenged the limits of what can and cannot be experimentally detected given the technological means available to the experimenter. Timing a falling body was probably just as challenging at the time of Galilei as the Müller and Peters experiment is today. Apart from the technological challenges cutting edge experimenters face at all times, the fundamental similarity between the two experiments is the requirement to create a laboratory system, that is to shield the experimental setup from environmental fluctuations and disturbances. In the case of the Müller and Peters experiment part of this "shielding" process is literally "shielding" the system from temperature fluctuations. The shielding process can be more indirect, built into the experimental setup. Galilei's polished ridges and parchment lined grooves are what "shields" his experimental setup from the environment.

Designing a laboratory system, however, is not only about shielding the system. The experimental representation of the idealised system can itself go a long way in increasing the accuracy of the operation. Galilei chose inclined planes rather than actual free falling bodies in order to increase the time of the fall, which in turn increased the relative precision of his measurements. In the Müller-Peters experiment, and indeed in the original MME, the experimental setup that measures differences rather than absolute values of the speed of light had a similar role of increasing precision by clever design choices.

The idea of a laboratory system is inextricably linked to the process by which theories in physics are arrived at. Experiments nearly always refer to a theory, in the sense that they set out to test a specific mathematical prediction from the model. With an insistence on mathematics comes the necessity to simplify radically, to idealise. Without ignoring nearly everything there is in the world, the mathematical theory would not be tractable. The physicist would not be able to analyse the equations that make up the model; maybe she would not even be able to write down the equations. This principle of idealisation, although convenient for the theoretician, leaves the experimental physicist with a big problem: How can she test the theory developed for an ideal world within the real world that is far from ideal? The answer is the laboratory system. An artificial highly shielded system designed to create artificial conditions that fit the prescriptions of a theory.

Not every experiment is as momentous as Galilei's inclined plane experiments. In many instances, and the Müller and Peters experiments is a case in point, experiments are not expected to falsify a particular theory; many experiments are incremental improvements to measurement techniques. The art of creating ever more sophisticated laboratory systems is often an aim in itself. Yet even in these cases, the justification of the experiment, why money and time has been spent on the experiment, will always be in reference to a specific theory, just as Müller and Peters point to the theory of special relativity. At the same time, probably nobody ever holds their breath and waits for the outcome of these empirical tests, ready to throw relativity over board.

Chapter 3

The science of Life

Was ist dir, Tor?
Ach, bist du dumm!
Bist doch weder Vogel noch Fuchs?
Siegfried, Act 1

Most would probably agree that physics was successful as a science; indeed most would agree that it still is. Yet, what precisely is it that indicates its success? Why do we think that the story of physics was a success? It seems that one important aspect of this success is its ability to explain natural phenomena. There is a practical and intellectual benefit to this. The moment we understand natural phenomena as such, we also liberate the human spirit from myth and superstition. At least some see it that way. Physics has, in a sense, provided us with a rationalist plot-line. Prescientific man explained natural phenomena by inventing entities and myths. So, for example, they explained lightening as a result of the ire of the gods, and solar eclipses indicated celestial struggles, and natural disasters were punishments sent from the heavens. Then came physics to set things right. Now *homo scientificus* knows that natural disasters just happen, celestial struggles, to the extent that they exist at all, do not manifest themselves through solar eclipses and that Gods have other ways to express their rage.

It is true, that physics has provided explanations. Yet, it is also true that explanations are not in and by themselves specific to physics or even scientific theories. Invoking the ire of Gods as a cause for lightning constitutes an explanation as well. Apparently, though, we seem to think that it is in some sense a worse explanation than that provided by physics. Why is that so?

Why do we think that recourses to the actions of the Gods is not a good strategy for explanations? One aspect is that many may not believe that Gods exist. This could be rectified and we can replace "Gods" by nature. The lightening is the ire of nature. Or one could say that nature has a natural tendency to produce lightening when there are thunderstorms. This certainly constitutes an explanation of some sort for the phenomenon, although, the reader would probably agree with me that none of those are very good. We could rationalise this as follows: These "explanations" are just made up stories that have no empirical or rational justification. They are quite pointless explanations.

Physical explanations are different. They are not just stories: they come as theories. We now understand that solar eclipses are the result of a peculiar alignment of sun, earth and moon; lightning is electric discharges in the atmosphere and earthquakes sudden shifts of the tectonic plates. Physics provides better explanations in the sense that we now understand the base phenomenon (electricity, motion of bodies in gravitational fields, plate tectonics) and from this base understanding we can derive the particular events. Physics generalises phenomena and explains them as instances of larger classes of events.

Physics has been extremely good at unifying phenomena in this way. We now know, thanks to physics, that light and electricity (and your microwave oven at home) are fundamentally related to one another. All of them are described by the basic formulas of electrodynamics, specifically the Maxwell equations. There is more to unification. The mathematical apparatus to describe light is in many ways the same as the one used to describe water waves or sound waves. In this sense physics has found that apparently different phenomena are amenable to very similar explanatory approaches. Intellectually this is tremendously satisfying, particularly when compared to explanations that need to invoke Gods.

Apart from their intellectual beauty, physical explanations are also practically useful. Probably the most important reason for the importance of physics in the contemporary world is its ability to spin off technological advances. Nuclear weapons are the most notable advance, but then there are also semi-conductor devices and computers, steam engines, mechanical devices of any sort, tele- and microscopes to mention but a few. They are all direct applications of scientific principles that have been established at some point. The importance of practically useful spin-offs must not be underestimated. These are, in the end, the results that motivates the public to spend money on research. Without them, science would have remained a hobby-activity with the social prestige of potato peeling.

Physics is also scientifically successful, to the extent that physics envy is a recognised condition among scientists. What sets physics apart from all other sciences is the integration of experimental research, mathematical analysis and creative hypothesis generation—the Galilean paradigm. Many legitimate fields of research have very different modi operandi. I like to think of the stereotypical biologist who is sitting somewhere in the field studying the behaviour of scarab beetles, or collecting plants and fungi for further study and classification. Maybe she follows a particular pack of wolves or tries to understand the behaviour of that rare and endangered bear species somewhere in northern Siberia. She probably spends half of her time in the field, then comes back to her office and describes what she has seen in books and articles using well-crafted illustrations. The kind of research these biologists are doing is certainly valuable and probably of more practical use than what many physicists do. At the same time, from a physics perspective it is reminiscent of pre-Galileian physics. It looks more like a proto-science than a science. When the field biologist observes rare snow foxes in Greenland they learn nothing about tigers in Nepal or bat-eared dogs in Tanzania. Each individual research report is isolated information, nothing like the unification of physical theories.

Some describe this type of science, maybe with a hint of ridicule, as *stamp collecting*. I am sure they know that this is fuelling an inferiority complex among life scientists.

One of the main problem biologists face is that organisms are hard to study. Depending on the species they may swim, fly, mate, search for food, sleep, fight. All this makes them very interesting, but they do not seem to share a unifying principle. The typical life histories of Siberian tigers and the red billed hornbill will be very different as are their ecological roles and contexts. Hence, the need for stamp collecting. To be fair to those field biologists who spent countless hours waiting and watching, certain general patterns did emerge. For example, fighting is more likely during the mating season and between males; mating often happens during spring; the size difference between male and female animals depends on the reproductive strategies and social organisation; and then there are many more rules that biologists have learned from studying many different particular examples. So stamp collecting led to a limited generalisation, but there are always exceptions, and many of them, so there is no hope of generalising this to a comprehensive theory, certainly not a mathematical theory on a par with Newton's laws. It seems that there are particulars to any species that are essential to this specific group of organisms. These particulars cannot be ignored without losing essential information. In this respect the situation is very different to the Galileian approach to science. With respect to their free-fall properties, Siberian tigers, ants, paperclips and teabags are all the same. Their difference in function, colour, shape and size is irrelevant. In the domain of stamp-collecting biology, every species has its particular behaviours, constraints and adaptations that distinguish them from any other species in relevant ways. The particulars count.

Moreover, what is worse, even if there was a general theory, it would be difficult to test in the lab. Take for example mating and fighting. These are important aspects of animal behaviour. They also depend crucially on their ecological context. Territorial fights need to be settled in an area where there is a territory to fight about. In a lab this can hardly be realised. When in captivity animals tend to alter their behaviour with respect to the wild; they often stop mating and they certainly will no longer be able to roam their territory. This means that, the moment we take the red billed hornbill or the arctic foxes out of their natural environment, we change the system that we are interested in. The research subjects alter their behaviour, by necessity. Hence any lab observations are at best of doubtful relevance for the wild case. Compare this now to the rolling balls of Galilei. We have no reason to assume that they fall any slower or faster just because they are in Galilei's cabinet.

Stamp collecting in biology is a respectable thing to do. Stamp collecting in other fields may not be so. It would be difficult to get a PhD awarded in experimental physics by simply going around and recording the world as it is, watching objects fall and describing how long it took before they hit the ground. This would not be acceptable. At the same time, this does not mean that field biology is not a useful thing to do, or not a science. This branch of biology is successful by at least some of the above criteria and certainly important. While it does not generate unified theories, it is practically useful in all sorts of ways. Both physics and field-biological stamp collecting are sciences. Yet, both seem to follow very different standards of rigour and different methodological approaches. This is a first hint at a general theme that we will explore in the rest of this book: There may not be one single scientific method that demarcates research from other ways to generate knowledge.

Replacing the what with the how

Even within biology itself there is not a unified way of doing things. Not all of biology has given up on the aspiration to develop a unified theory of living systems, or at least something similar. Research in biosciences is not all only about collecting field data. A dominant direction in contemporary biology is concerned not so much with *what* biological systems are doing, but *how* they function internally. Such a theory could take the shape of a theoretical system that unifies all the individual "stamps" in the collection into a coherent – maybe even mathematical – system.

There are many different potential approaches to address "how-questions" in biology. One approach to explain empirical data in biology is the theory of evolution. The theory of evolution is particularly concerned with the question about how organisms change over time. Needless to say that it has had some success in doing so, and it has certainly had great intellectual impact. In many ways the theory of evolution is a theoretical superstructure over all of biology. However, it still lacks many of the features of physical theories. It is barely testable in the laboratory, it is not predictive, at least not in a quantitative sense. Evolution is not very specific. It gives a good overall view of how organisms change, but it does not tell us much about what is going to happen to Johnny the Bonobo specifically. The most deplored shortcoming, however, is the lack of mathematics in evolution. There are theoretical biologists who use equations and formulas to study evolution, but this community remains a small part of the field as a whole. Many feel that a scientific theory without mathematics is a bit like soup without salt...not very enjoyable.

Apart from the "how did species X come about" type of question evolutionary biologists like to ask, biologists also wonder about the mechanistic details of organisms. So, how does an organism work? To understand the kinds of questions that biologists are asking there, let us first recall the way that physicists would describe events: Consider as an example a stone that is thrown into the air. If this is done here on Earth, then the stone will fall back down to the ground. A physicist would explain this by postulating a gravitational field exerting a downwards force onto the stone. She would then further assume an initial momentum (or speed) of the stone in the upwards direction. The gravitational field decelerates the upwards trajectory of the stone. As a result it will eventually, for a brief moment, come to a halt mid-air, and start to fall back until it hits the ground at some point. If this is explained within a Galileian framework with the appropriate formulas and laws, then this establishes an explanation of the behaviour of the stone.

The idea of this explanation is that the stone is itself rather passive. It does not do much except being accelerated and decelerated by external forces. Indeed a physicist would not even take into account that it is a stone, or that it is thrown on Earth. All that matters is that it is a mass point and that it is in a gravitational field and that it has an initial momentum. Everything else about the stone is irrelevant.

Biology seems to require different kinds of explanations; this is at least what seems to be the case at first sight. Organisms are not mere passive objects in the environments but they are agents themselves. They actively do things, and do them in ways that seem very difficult to predict. For example, they reproduce. Take one and another one and wait and when the conditions are right you may end up with three (sometimes you do not even need two to get another one). They also eat one another, fight, dig holes into the ground and so on. This agency and internal behaviour of biological systems makes it difficult to force them into a coherent and precise mathematical theory. It makes little sense to model organisms as particles without internal properties. It is pointless to abstract them to mass points that are subjected to the forces of external fields. It would be a futile exercise to try to shoehorn the behaviour of the Svalbard reindeer into a mathematical formula.

Even if we had a comprehensive mathematical theory of how organisms work, it would not be very satisfying. We would still want to know: Where does the agency of biological systems come from? How come that there are entities in the world that are able to display autonomous behaviours? What is it that gives some objects agency (centipedes) when others do not have it (neutrinos)?

Perhaps it *is* possible to understand biological systems in terms of non-biological physical interactions, if we just look at the correct level of description. To understand what this could mean it is useful to compare living systems with machines. In many respects machines are not dissimilar from living systems. Writers of literary fiction have long been interested in machines that resemble living systems and emulate some of their functions. Mary Shelley's Victor Frankenstein creates an artificial human, with disastrous consequences. Not quite as skilled as Frankenstein, but still impressive enough to fool the protagonist of Jacques Offenbach's opera "Les Contes d'Hoffmann" is the physicist Spalanzani. He creates a life-sized doll called Olympia that enchants Hoffmann who falls in love with this golem. Rather amusingly, in the opera Olympia is shown as a completely mechanical creation. Spalanzani needs to rewind her several times, interrupting Olympia's aria ("Les oiseaux dans la charmille"). Despite all, Hoffman is fooled by the physicist's creation, in part because he wears special rose tinted spectacles. A more recent addition to life-like artefacts is Commander Data from the series Star Trek (second generation). Although he is just some well-disguised electronic circuitry, at a first glance he looks and behaves like a man made of flesh and blood. Commander Data is perhaps a bit of an odd-ball, but his eccentricity aside, he shows the same kind of agency that one normally associates with real humans. At the same time, he is nothing but cleverly designed, human engineered electronic circuitry; hence, he is nothing but applied physics and thus entirely reducible to the principles of a Galilean science.

These life-like creations are all figments of writers' imaginations, but these days they do find their real life counterparts.

Research into robots is active and will one day, probably, result in a finished product that looks and acts the part, at least to some extent. One of the cuter examples is AIBO, a robot dog made by Sony. It is programmed to show the agency of dogs, growing up from a little puppy to an adult dog. It understands human commands and may even disobey them if it "does not feel" like following. So, it shows agency.

More mundane everyday machines also show agency at least to some extent. Cars are examples. True, most of them need an operator, a driver, to harness their potential agency, but still they show behaviour as long as they are prodded in the right way. These days cars can even show a bit more autonomy. There are now examples of fully autonomous cars that can safely navigate the roads on their own. This has been impressively demonstrated by the Visilab Viac challenge (http://viac.vislab.it). In 2010 a project funded by the European Union demonstrated an autonomous vehicle that could complete a 13000km trip from Milano in Italy to arrive in time for the 2010 World Expo in Shanghai in China. With minimal human intervention the car travelled the entire distance on its own. For safety there were always operators at hand to take over in emergency situations and to make decisions where to go when the maps were of low quality. The vehicle was not fully autonomous and needed to get fuel from its human carers and it also needed some directions as to where to go in the sense that it did not have the independent will to go to China, but had to be programmed to do so. Even so, it is still fair to say that the Visilab vehicle showed some agency. Not as impressive as Commander Data, but it is real and in the here and now. We know that this car is entirely a human creation, built from scratch using materials and components that themselves do not show agency, purely physical parts, each of which is understood based on the laws of physics. None of the components has any agency, yet when assembled in the right way, collectively they do.

What is it now, that gives the car agency? The good thing about the Visilab car is that we can actually get an answer to this question, as long as we ask the right person. The best way to get the information we want is to ask an engineer to explain to us the internal mechanisms of the vehicle. Provided she is inclined to help us and has some pedagogical skills, she would probably first explain the individual high level parts of the vehicle. Assuming that we know how a car works, she may start outlining the sensor system of the car and then point to the artificial intelligence core that steers the autonomous vehicle. She would explain to us the basic principles of the algorithm that powers the artificial intelligence and processes the information received from the various sensors mounted on the vehicle. How this is combined with GPS location data to determine both the route of the vehicle and to make sure it adjusts itself to traffic and road conditions. We would perhaps be inclined to ask in detail about the algorithms that are used and how the parameters of these algorithms are set. And again, as long as we do not over-stretch the engineer's patience, we will eventually get our answers and should be able to build up a good understanding of how the Visilab autonomous vehicle works at a high level.

Once we feel confident about the individual modules that implement the autonomous behaviour, we could start to ask more detailed questions, about the car itself, but also about how the sensors and the individual parts of the car work. We could enquire about the properties of the rubber in the tyres, the pistons of the engine, the cogwheels in the gearbox and so on. Eventually we would give up because the amount of information would become a bit too much to digest. Still we could be confident that with some more persistence we would eventually be able to find out about every conceivable question we may have about the functionality of the car.

The relevant point for us is that the agency of a machine can be explained in terms of its constituent parts, at least in the sense that there is somewhere an engineer who can give a detailed explanation of any part, or failing that a physicist who understands the physics behind the specific effect that has been exploited in the engineering. In the Visilab car agency emerges from parts that have no agency, and there is absolutely no mystery about that. In some sense, the car is *just* a collection of parts each of which can be described in terms of physical laws. They are physical objects that are completely passive in the environment. These components do not move about by themselves. Unless moved they would simply remain where they are until they decayed eventually. Collectively, however, they can drive to Shanghai.

Based on the idea that machines can be explained entirely in terms of their constituent parts, the question arises whether or not the same holds true for biological systems. Is there a way to explain bacteria, plants, animals or any other type of living system purely in terms of physical interactions? Is it true that they, like the Visilab car, are nothing but the interactions of their dead parts? Can we reduce biology to a specialised branch of physics? Can we use this approach to incorporate life sciences into the Galileian family of sciences, probably ultimately dispensing with the need for stamp collecting altogether?

This reductionist programme of biology is an exciting possibility. From a scientific point of view, life is a somewhat vexing property. It is a bit mysterious. Indeed for a long time, scientists thought that living systems are somehow fundamentally different from non-living ones. We no longer believe that, but then again, it still is a big problem to understand how life arises from non-life. The prospect of reducing cells to their underlying physics is alluring. The fact that it works with the car is encouraging, but there is no guarantee that the same strategy will carry over to organisms which seem to be much more complex. At the very least, however, the Visilab car gives us an analogy to work with, a methodological handle that allows us to start to think about a science of living systems that goes beyond mere stamp collecting.

Beyond stamp collecting

Trying to reverse-engineer biological systems is big business in academia. Governments and private corporations invest much money into this type of research. Bioscience has also become an intellectual force of formidable might. Over the past 60 years or so biology has graduated from a field with not much more on offer than countless numbers of stamp collections, to an academic giant that is now at least as respectful as nuclear physics ever has been. There has never been a biological Galilei, a single founder of the new biology. The birth of the new biology was the collective achievement of a handful of scientists who over a period of time made key discoveries about how cells function developing a new paradigm for biological research in the process. What they all had in common was a strong dissatisfaction with simply describing what organisms are doing. They challenged themselves to understand how organisms were doing what they are doing. One of the most influential ones amongst those pioneers was the French biologist and Nobel Laureate Jacques Monod.

Monod's foundational work in microbiology defined biological research for decades to come. Indeed the spirit of Monod pervades the pages of modern day textbooks in biology. His idea for the new biology was precisely to abandon stamp collecting in favour of a systematic study of the underlying mechanisms of organismic agency. Monod wanted to explain life in physical terms. To this day, biology has not been reduced to physics. The language of microbiologists, their methodologies and also the tools they use are distinctly biological. No physicist would find herself at home in this world. Still, the conventions that have since been established, the rules defining what is good biology and what is a valid question clearly reflect the quest to understand biological systems at a microscopic scale. Biologists want to treat living systems as physical objects. Life is now no longer explained in terms of an *élan vital*. Instead, biologists are starting to describe organisms as mechanisms consisting of components that themselves can be reduced to elementary physical laws. The science is still far away from reducing *E.coli* to Newton's laws, but the biologists have embarked on this journey.

In practice there are a number of problems when explaining living systems and their agency. One of the main difficulties is that it is unclear what particular aspect should be explained. Compared to, say, a car that has a particular function, and a well defined behaviour, biological systems tend to be vastly more diverse. The function of the Visilab car is to drive autonomously from Milano to Shanghai. This is what it is doing, nothing else. So, if you wonder how it works, you may be interested in how the autonomous navigation works, or how the car proper works. Those are the modules that need to be understood in order to understand how the car functions.

Living systems are much more diverse in terms of their behaviours. For this reason, it is so much harder to translate big questions such as "How do organisms work?" into concrete and feasible research problems that can be addressed in the laboratory. In order to be able to make any tangible progress a scientist needs to be able to formulate hypotheses and design experiments. "What is the physical basis of the agency of organisms?" is not a question that can be directly translated into a testable hypothesis.

One of the first steps to narrowing down the general question to a specific one is to choose a particular organism. The scientist needs to concentrate, at least at first, on a particular species. Understand the particular first, then try to generalise. From a purely practical perspective it makes sense to choose organisms that can be accommodated within a laboratory environment and that are easy to keep and reproduce. Rabbits or rats fit the bill and are therefore popular laboratory animals. For some of the most fundamental questions in biology there is an even better choice: Bacteria.

Bacteria are very diverse. The various species are able to populate a truly impressive range of habitats. Bacteria have been found in the boiling deep sea-vents, but also frozen into the apparently life-less world of Antarctic ice. We tend to think of bacteria as living fossils, primitive life forms that have long been superseded by higher animals and are mere remnants. The truth is that bacteria are highly adapted and continuously evolving molecular machines that are extremely adept at invading new ecological niches and maintaining those that they have already colonised. Compared to rabbits and mice bacteria are organisationally rather simple creatures. This is, however, not to say that they are not well adapted.

Bacteria are unicellular organisms whereas higher animals such as rabbits consist of a very large number of cells. One can think of a bacterium as a compartment, a bag filled with water and chemicals. The bag itself is a so-called lipid-bilayer. Such lipids are small, but very useful molecules. They are rod-shaped and the magic of these rods is that each of its ends interacts slightly differently with water. One end is so-called "hydrophilic" or water loving. It is not really water loving, but it carries a small electric charge and thus mixes well with the water molecules that equally carry a small electric charge. The other end of the rod does not have such a charge, but is electrically neutral. The significance of this can be easily seen in computer simulations. If the conditions are right and one throws these rod-shaped molecules into a bottle of water then they will quite naturally start to form compartments, or bags. These structures are called micelles and they are not that dissimilar to the membranes of real bacteria.

In bacteria, this lipid bilayer, the cell membrane, has a simple function. It separates the inside of the cell from the outside. This separation enables the cell to maintain a highly specialised chemical cocktail. The chemical composition of the inside of the cell is very different from the sourrounding environment. The membrane is not entirely impermeable though. It selectively lets molecules from the environment enter the cell, and it can also excrete chemicals from the inside, releasing them into the environment. There is thus a constant, albeit tightly controlled exchange of matter between the cell and its environment.

The chemical cocktail that the cell maintains has a remarkable property. The mix within the cell is just right so that when it is fuelled from the outside with the right sort of material, the cocktail can make more of itself. Not only that. The cocktail can also make more of the bag that contains it. This is the essence of reproduction in bacteria. Indeed, this is the essence of bacterial life. Creating more of themselves.

The most important types of molecules in this cocktail are so-called proteins. Proteins are very large molecules often hundreds of times larger than, say, an oxygen atom. They can be thought of as linear chains of amino acids, another type of molecule. In organisms there are 20 different types of amino acids. Their particular sequence determines how the protein folds up into a 3D structure and thus in turn its chemical properties.

Depending on the amino-acid sequence, the proteins have a wide variety of different properties and capabilities. Some of them are enzymes, that is they facilitate specific chemical reactions. This is a crucial task for life, but maybe not that spectacular. Some types of proteins found in cells have the unique task of destroying or modifying other molecules, often other proteins in the cocktail. In higher cells this is often done by transport. One protein attaches to another one and pulls it into the molecular scrapyard within the cell. Some of the proteins are capable of cutting and remodelling molecules. There are proteins that act as motors and propellers. Bacteria use these in order to be able to actively move through their environment following nutrient gradients. Then there are still other proteins that form openings in the cell surface to allow exchange of chemicals with the environment. However, these proteins do not let anybody in. Like the bouncers in posh London night clubs they are very selective in deciding which molecules get in and which ones have to stay out. With this wide variety of properties, individual proteins are only different in how many amino acids they have and how these amino acids are arranged on the chain. Once an amino-acid sequence is specified, so are the properties of the protein. In a way, the entire variety of life we see on Earth boils down to different ways of stringing together 20 amino acids in various combinations.

In many ways proteins can be seen as the bridge between the kingdoms of physics and life sciences. Some of the more complicated proteins, the motor-proteins or the restriction enzymes that cut other molecules, almost seem to show agency. At the same time, we know that these proteins and their function entirely depends on their shape. The shape again only depends on their sequence. Except for the very smallest proteins, we are currently not able to calculate the shape. Still, one day this will be possible. There is the dangling carrot at the end of the stick that promises us to link the quasi-agency of proteins with the dead world of amino-acids.

Another type of molecule in the cocktail is DNA. It is the hereditary information of the cell, its genome. Sometimes it is said that it is the blueprint of the organism. This is perhaps a wrong analogy. Still, the DNA contains crucial information required to "build" organisms. As a molecule, the DNA is special because it exists only in a few copies per cell. In bacteria there is often only a single DNA molecule. This single molecule contains a description of all the proteins that can be found within the cell. It is an amazing thought. The entire life of the bacterium depends on one single molecule. Just like proteins, the DNA is a chain of smaller components. However, when proteins are sequences of amino acids, the DNA is a sequence of nucleotides. Compared to proteins, the DNA is a very long chain, consisting of millions of nucleotides. A protein is at most a few thousand amino acids long. Indeed, the DNA has to be longer, because its sequence of nucleotides describes, or "codes" for each of the proteins in the cocktail.

One way to think of the DNA is to picture it as a string of beads, where each bead represents a gene which in turn codes for a protein. In order to produce a new protein a particular section of the DNA sequence that represents the right "bead" must be read and decoded. Biologists understand now that this is a multi-step process. First a specific reading protein attaches to the correct sequence of the DNA that codes for the protein to be expressed. It then starts to scan the nucleotide sequence and produces an intermediate molecular message. This message in itself is not a protein, but rather again a nucleotide sequence, similar to the DNA, but not identical. The main difference is that this intermediate code is much shorter and only contains the description for a single protein. We can think of it as floating around within the cell volume. The cell also contains a class of scanning molecules, little molecular reading heads that can decode the molecular messages and build proteins, one amino acid at a time.

The survival of the cell depends on it producing enough proteins, and in the right proportions. Ultimately, cell life, particularly bacterial cell life is about making more of itself. Yet, this ultimate end cannot always be achieved by just production production. In times of plenty, the cell must crank up its life cycle and produce more of itself. However, in times of scarcity it is better to wait until more nutrient is available again. Producing proteins is not free for the cell but requires a substantial energy input. The wise cell therefore regulates when it produces and what it produces.

Among the fastest reproducers are bacteria. When conditions

are right they can produce enough proteins to grow and eventually divide several times within an hour. For the researcher this is very convenient because it allows her to see straight away, or nearly straight away, the effects of any changes she made to the cell. Without having to wait for unduly long times she can test how a new feeding regime impacts on the cell or how a modification of the organism impacts on its behaviour.

All this makes bacteria the perfect lab-rats. They are small, easy to grow, reproduce fast and they are still in touch with their chemistry. Compared to animals or plants or even simple fungi, these bacteria have very little internal structure. They are protein-factories in a bag, a system of interacting bio-molecules on the one hand, but at the same time organisms with agency. There is no other organism in which the physical and the biological are so close together. In this sense, bacteria are our best bet to understand biology in terms of physics.

The growth of biological knowledge

Unlike ourselves, bacteria do not normally reproduce sexually. For a new bacterial cell to come into existence there is no need for two parent cells. One is enough. To produce offspring, bacteria simply divide. Once a cell has made enough of its own chemical cocktail and increased the size of its bag sufficiently, it simply splits into two cells, one of which is the offspring and the other one the parent.

This simple pattern of reproduction has interesting mathematical consequences. Let us assume that there is a single cell in an environment that has plenty of nutrient in it, so it finds ideal conditions for growth. Under such perfect conditions the bacteria can achieve a generation time of about 20 minutes. This means that if there is a single cell to begin with, then after twenty minutes there will be 2 cells. After another twenty minutes each of those two cells will again have divided, which means that there are now 4 cells. Then again, all of these 4 cells divide, and there are 8 cells, then 16, then 32, 64, 128 and so on. Every twenty minutes the number of cells doubles. At first this doubling dynamics is rather innocent. It takes about 2 hours to get to 64 cells. Hardly impressive. However, after about 6.6 hours there are 1048576 (over 1 million) cells. This is now a little bit impressive. After 12 hours there are 1099511627776 (a trillion) cells, which slowly turns from impressive to scary.

This growth dynamics is called *exponential growth*. What makes it so special, or interesting, or scary is that it is not a simple growth or acceleration as we know it from our daily lives. It is a growth with accelerated acceleration. The more cells there are the faster the increase of the cell number. During the first hours only a few new bacteria are created, but after 20 hours it only takes 20 minutes to create another trillion cells. This is the ultimate run-away process.

To grasp this, think of the analogy of a car. A normal car accelerates from 0 to 100 km/h in, say, 10 seconds. This is the normal kind of acceleration a typical street car can achieve. Most cars can drive much faster than 100 km/h. Some of the very fast ones can probably reach 200 km/h. However, getting from 100 km/h to 200 km/h will take more than 10 seconds. Probably 30 seconds or a minute, depending on the car. In any case, the faster you go, the longer it takes to further increase the speed. We all know this. Bacterial growth is very different in this respect. A bacterial car, if you will, accelerates faster the faster it drives. So, it would accelerate from 100 km/h to 200 km/h not in 30 seconds, but in, say, 5 seconds. It would then continue to speed up and reach 400 km/h in 2 and a half seconds, just to get to 800 km/h in a further 1 second and a bit later. To sit in such a car would be a truly scary experience.

Exponential growth is very common in physical, life and social sciences. Take economics: The GDP of the United States grows by about 3% annually. This is not quite the doubling dynamics seen in bacteria, but the principle is the same. Each year the 3% share by which the economy grows will be bigger than the year before. Even if the percentage growth remains the same over the years, the absolute amount by which it grows increases each year, simply because 100% corresponds to a bigger cake. In 2010 the GDP of the USA was 14526550 million dollars. If it grows at 3% then this means that it adds another 435796 millions of dollars to its GDP. So, just the growth bit is somewhere between the size of the economy of Saudi Arabia which was ranked the 23rd largest economy in 2010 and the Republic of China which was 24th. Assuming the same growth of 3% again in the following year, it would grow by a further 448870 million dollars. Now, this increase is nearly the size of the entire Swiss economy, which was ranked 19 in 2010. So the chunks by which exponentially growing economies get bigger every year are quite substantial.

Prices show the same exponential increase. This is the all too familiar phenomenon of inflation. The moderate inflation rates that are common in the developed world are benign in the short run, but over time will show precisely the kind of runaway dynamics of bacterial populations, only on a slower scale. If everything gets more expensive by about 3% every year, then this is a small increase at first, but as time goes on the absolute increases in prices year on year keep getting bigger and bigger and bigger. Just as in the bacterial population.

Mathematically the behaviour of growing bacterial colonies is extreme. If bacterial colonies really grew exponentially for a sustained period of time, then after not too long there would be more bacteria in the colony than atoms in the universe. This is clearly not possible. Sooner rather than later there has to be a show stopper. Incidentally, this is not only true for the exponential dynamics in bacterial colonies, but applies to any type of exponential growth, including that of economies. What stops bacteria from multiplying is the depletion of the nutrients they require to fuel their growth. No more food means no more protein. No more protein means that the whole production engine that drives bacterial growth will start to stutter and ultimately come to a halt when all nutrients are exhausted.

In reality nutrients can become exhausted quite rapidly, often stopping exponential growth before it even started. For example, when bacteria divide offspring are usually created close to the parent cell. When the bacteria do not move, or only move slowly, then this leads to clumps of cells. Part of the local environment will become very crowded and cells will start to compete for resources with other cells in the immediate neighbourhood. This is a bit like in the supermarket in the evening when the food that nears the end of its shelf life is priced down. There are incredibly good deals to be had. Bread at a fraction of the normal price and desserts or fruits for nearly nothing. Typically, the supermarket is quite empty at that time, but around the shelf with the marked down goods are you can still find a little crowd fighting to get what is there. The same crowds develop in bacteria when offspring is created close to the parent. Just as the bargain hunters compete for cheap yoghurt, so will the bacteria compete for nutrients in the immediate environment. Such local competition can stop exponential growth in its tracks, even if there would be plenty of food in other places in the environment.

To understand let us consider the case of Amelia who likes to go to the forest to forage for mushrooms. Assume that the weather had been just right for mushrooms to grow and there are plenty of them around. She would not have any difficulty finding mushrooms. All she needs to do is to walk through the forest, pick whatever she finds and go home when she is tired or when she thinks she had enough. She will be back with filled bags, mushrooms for days to come. Maybe she even has enough to go and sell some to friends. As far as Amelia is concerned, the amount of mushroom in the forest is practically unlimited.

Now imagine that Amelia is not going to the forest by herself, but instead with a large group of friends. Since this is a social outing, assume that the main purpose of the tour through the forest is to chat, to spend some time with friends; foraging is a nice excuse to get together, but mostly they come to the forest to catch up with one another. So the group stays together during most of the foraging trip. Say, they stay within a radius of 10m. We again assume that there are plenty of mushrooms in the forest, but the dynamic is a bit different now compared to Amelia coming on her own. If all of the members of the group actually look for mushrooms and pick any they spot, then Amelia will suddenly find it much harder to fill her bags. The reason is that only the person who sees a mushroom first will be able to take it. The rest of the group gets nothing. Provided the group moves through the forest, then collectively they will still be able to collect as many mushrooms as they like, but clearly, for each one of them it will take much longer to fill their bags compared to single-person foraging. The time required to find a given amount of mushrooms depends directly on the size of the group. If the group size is 10, then they need to stay 10 times longer in the forest compared to single foraging; if she goes with 20 friends, then for every hour solo foraging she has to spend about 20 hours in the forest. The reason for this is intuitively clear: The members of the group deplete the resources locally, they pick each others' mushrooms. Therefore a much wider area must be sampled to collect the same amount of mushroom as a single forager.

The linear relationship between group size and foraging time depends on the assumption that the group actually moves about. If they simply go to the forest and stay put at one place, then they will clearly only find the mushrooms that grow in this particular area. No matter how long they stay there, they will never be able to fill their bags (we ignore the possibility of mushrooms growing while they are in the forest). This is the case where the local depletion of nutrients is toughest. Only the friends who find the mushrooms first will have anything. Most will probably never find any mushrooms and come home with completely empty bags. This scenario is also the one where the global resources of the forest are least exploited, because most of the mushrooms of the forest are simply left untouched, never to become a mushroom soup.

Let us now consider another variation of the group moving. Consider the case where the friends started an argument with one another and prefer not to spend the day together. Then Amelia and her friends spread out over a larger area of the forest so that they do not need to talk to one another. For simplicity, let us assume that they take random positions in the forest. They forage more or less independently. If the forest is large enough compared to the size of the group, then they will rarely meet, if at all. Consequently, there will be hardly any competition between the friends. Amelia, even though all her friends are in the forest with her, will again be able to fill her bags within the shortest time, as though she was alone all on her own. This tells us that clusters of mushroom pickers are bad for the foraging efficiency. If you want to go home with many mushrooms in a short time period you should either go by yourself, or if you insist on group outings, spread out and do not stay close to other mushroom pickers.

Bacteria face exactly the same challenges. If they do not spread, then they will start to compete with one another for nutrients in their immediate environment. If the bacteria do not spread out but cluster in specific spots, exponential growth will stop before it gets started properly. In their natural environments bacteria will only rarely find the conditions that allow them to avoid spreading. For example, our guts are heavily colonised by bacteria. The bacteria live there essentially on a surface, sometimes they are even attached to the host cells. In any case, their scope for movement is very limited. The offspring of bacteria is likely to remain close to the parent cell and competes with it for nutrients, just as Amelia competes with her friends for mushrooms when they do not move very fast. Most of the time there will be no exponential growth of bacteria in their natural environments. This is a good thing in most contexts, as we would not want to be overrun by them.

This natural environment of the bacteria is not convenient for the scientist who studies bacteria. Dealing with the conditions of limited bacterial growth is methodologically difficult. One problem is the reduced reproduction rate compared to exponential growth. If a bacterium can divide every 20 minutes under ideal conditions, in real conditions it will rarely be that fast. For the scientist this means longer waits and fewer cells to experiment with. Apart from this convenience issue, another problem is that the individual members of the population find different environments. Those at the edges of the cell clump find plenty of nutrients and those at the centre find no nutrients at all. Experimentalists do not like this. Ultimately they want to understand how individual cells react to specific environments and therefore depend on all cells "seeing" the same environment.

Clumps of cells are also very difficult to model mathematically. In exponentially growing cells the modeller can ignore most aspects of the environment. All she needs to worry about is the total amount of nutrient that the cells find and how many cells there are overall. When there are clusters of cells that deplete only parts of the environment then suddenly the mathematics gets much harder and the modeller needs to account for the spatial distribution of the remaining nutrient. This is difficult and most of all it is against the idealising spirit of Galilei and should be avoided. In short: Local competition of bacteria is a large source of uncertainty and hence a headache for a scientist who wishes to control her system as much as possible.

Monod understood this very well. Real bacteria in real environments were no good to him. He could not effectively find out about them, just as Galilei could not effectively find out about the free fall by looking at how leaves fall from the trees. In order to make any progress he had to idealise the system. He clearly could not idealise the cells themselves, at least not at first. So, instead, he idealised the behaviour of the cells by creating an environment where the cells behaved in a way that made them amenable to a thorough investigation. Specifically, he wanted to have them under well-defined conditions where every cell is exposed to the same kind of environment and reproduces rapidly. In other words, he wanted the cells to be spread out in the forest, so to speak, and prevent them from competing for nutrients locally. A simple way to achieve this is to suspend them in a liquid with nutrients, a broth, and continuously stir them. The effect is that the individual cells are at random places in their broth. Monod describes how this should be done:

Although the methods of culture will vary according to the problems investigated, certain general requirements must be met in any case. The most important one is that the cultures should be constantly mixed, homogenized, and in equilibrium with the gas phase. This is achieved either by shaking or by bubbling air (or other gas mixtures) or both. Bubbling is often found inefficient. unless very vigorous, when it may provoke foaming which should be avoided. Slow rocking of a thin layer of liquid is the simplest and probably the best procedure. (Monod, 1949, p. 378)

In this quote Monod gives very detailed instructions on how to prepare and treat the broth in order to get the desired behaviour. Maybe this reminds the reader of the quote by Galilei where he describes how to polish the balls and the ridges so as to minimise friction. Monod does not try to reduce friction here, but fundamentally his intention was very similar to Galilei. He wanted to create a laboratory system, an experimental set-up that creates idealised conditions and screens the system from external and internal disturbances. In that sense, this quote is a witness of how to do Galileian science in biology. It is a testimony of the (then) newly-found ambition to leave behind stamp collecting and turn biology into something that works more like physics.

Mixing of the bacterial culture can prevent the local depletion of nutrients and ensure exponential growth, but we have already seen that exponential growth cannot be sustained for too long in the real world, certainly not in bacterial cultures. A limiting factor in laboratory experiments is nutrient depletion. At some point, the cells have converted all the food there was into biomass, and there is nothing left to grow on. There are no more mushrooms in the forest so to speak. From an experimental point of view the limits to exponential growth are not necessarily a problem. For certain types of questions, however, the change of the chemical composition of the broth in which the cells are suspended could have unpredictable consequences on measurement results. Again, this needs to be excluded as much as possible, if one wishes to do science the Galilean way. Monod suggests the following:

As a culture grows, the conditions in the medium alter in a largely uncontrollable and unknown way. Therefore, the observations should be performed while the departure from initial conditions may still be considered insignificant. (Monod, 1949, p. 378)

These concerns about reproducibility and laboratory control are yet another sign that Monod is trying to do biology the physics way. The shadows of Galilei are unmistakable in Monod's instructions above: Take the simplest possible organism, take it out of its natural environment, screen it from all environmental disturbances and subject the organisms to the controlled environment of the laboratory.

For his early experiments Monod chose the bacterium *Escherichia coli* (which is commonly referred to by its nickname *E.coli*). Since the days of Monod this species has developed into one of the workhorses of microbiology and is probably the best researched organism. The underlying reason the bacterium is so popular is partly historical, partly functional (it grows well) and partly due to the fact that *E.coli* is also important from a medical perspective. Strains of the bacterium live within our guts all the time. It has been estimated that an average human colon contains about 1 kg of bacteria, some of which are *E.coli* bacteria. These strains that live within us are not doing any harm and happily coexist with us. However, *E.coli* is a major cause for urinary tract infections, a disease that can become chronic. There are also strains of *E.coli* that can cause severe food poisoning. Understanding *E.coli* is therefore not only of scientific interest, but also of practical use.

Monod and diauxie

With all these preliminaries in place, it was time for Monod to get going. He started by growing cells exponentially. The experiment to produce the growth curve in figure 3.1 is very simple. It shows the amount of bacteria on the vertical axis as a function of time at the horizontal axis. For his experiments Monod prepared a solution with a suitable nutrient and then added a small number of bacteria to the solution. Mixing, stirring and aerating as described above he would be able to generate exponential growth. Indeed, the upper graph is a straight line, which indicates that the cells are indeed growing exponentially. The lower graph shows the results of a different experiment, although under nearly identical conditions. To produce this second graph Monod again prepared a broth, again he added nutrient and again he added a given amount of seed bacteria into the broth. The suspension was again stirred and aerated in the same way. He then expected to see exponential growth again—but he did not. At least not at first. The resulting growth dynamics was somewhat different. Instead of giving the straight line one would expect for exponential growth, the graph is curved at first. This shows that the growth was slower than what we would expect from exponential growth. It seems that there is something that held the cells back at the beginning. The bacteria are not dividing as fast as they should.

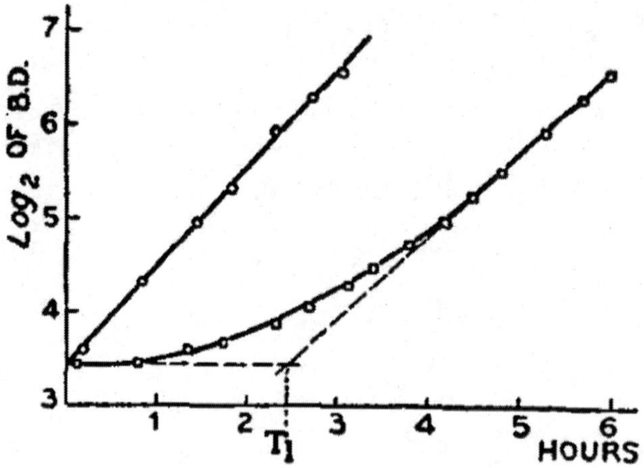

Figure 3.1: The growth of bacterial cells as observed by Jaques Monod (Monod, 1949, figure 7). (Permission obtained)

As it turned out, Monod found this kind of poor growth quite frequently in his experiments. So frequently indeed that he gave it a name. He called this initial period when bacteria are growing slowly the *lag phase*. To get a clue of what causes this lag phase Monod took a closer look at the details of the experiments that produced these curves. The two growth curves describe growth processes happening under nearly identical conditions, except for one detail: The straight curve represents a colony of cells grown on *glucose*. This is a type of sugar that bacteria can use as a nutrient. The other curve that shows the lag phase is grown on a different type of sugar, namely *xylose*.

Since this is the only difference between the two cultures, Monod rightly hypothesised that the difference in nutrients is the root cause for the difference in behaviour. Whatever the cause for the lag phase may be, it only seems to affect the population fed on xylose, but not the one that is in a glucose solution. One may be inclined to think that xylose is somewhat inferior and cannot give rise to exponential growth. This cannot be the case though, because eventually the xylose population also shows the characteristic straight line that indicates exponential growth. It is very puzzling!

Another classic set of experiments that Monod performed in the 1940s helped to shed some light on this mystery. These experiments were very similar to the above described work. He grew bacteria on different nutrient sources, as above, but with one crucial difference: Instead of having two separate experiments in two "pots" with one nutrient source each, he prepared one solution that contained two different nutrient sources in the same pot. The effect was that the bacteria were now exposed to a mixture of nutrients. So, one pot had two nutrients, rather than two pots having one nutrient each.

Probably Monod did not think that adding two nutrients together in one broth would make much of a difference. It is only reasonable to expect there to be exponential growth again. Maybe there will be a lag phase as well. In short, one pot with two nutrients is most likely the same as two pots with one nutrient each.

Here now, Monod was in for a surprise. Two nutrients in one pot were different. Monod suddenly found that there was a lag phase, not only at the beginning of the experiments. This would be unsurprising. Now he also found a lag phase in the middle of the experiments. This is very odd. The cells seemed to grow happily, dividing nicely following the exponential law. Then suddenly, they entered a lag phase, seemingly without good reason. When this was over they then resumed exponential growth again.

To understand the root causes of this odd behaviour, let us consider an analogous case: A computer science convention. Computer programmers are famous for their love of pizza. Therefore, if they are confronted with a choice between pizza and, say, a carrot with a humous dip they will invariably choose the pizza, only resorting to the vegetable option when they have no way of getting pizza. Let us now imagine that at one of those countless computer science conferences that go on every year the lunch buffet had exactly two choices, namely carrots with humous and pizza. Knowing the particular preference of the participants, we would predict that the pizza would be eaten first until there is none left. The carrots with humous would remain untouched as long as there is pizza. Once all the pizza is gone, the delegates would probably eat nothing for a while, hoping that somebody brings out more pizza. Only when they lose hope (and are hungry enough) would they grudgingly turn to the carrots.

If the bacteria have a similarly strong preference for one nutrient over the other, then this could explain Monod's findings. Imagine that the cells prefer one of the nutrients in the pot over the other and exclusively take up the one they prefer. They only turn to the less preferable one if they have no more choice. One way to explain the lag phase is to see it as the period of indecision: When the preferred nutrient has run out the bacteria are not sure about what to do next. Should they wait for more pizza, or just start eating the carrots? This is a bit fuzzy for the time being, but the basic idea can be tested experimentally. If the lag phase really has to do with the preference for one sugar over another one, then this means that the length of each of the two exponential phases depends on how much of each nutrient source is available, just as the time before the first programmer starts to eat a carrot depends on how much pizza was provided by the organisers of the convention.

Monod tested this idea by repeating the experiments with different amounts of nutrient. And, indeed, he found that the length of the individual exponential growth runs depends on the amount of each nutrient he provided. Figure 3.2 shows Monod's original results. The more glucose Monod used the longer the first growth-phase. This is a strong indication that *diauxic growth*, as he called the growth with lag phase in the middle, is a case of pizza versus carrot (and incidentally also that glucose is the pizza equivalent for bacteria).

This observation of diauxic growth gave Monod also a new interpretation of the lag phase. The underlying cause for the lag phase seemed to be connected to the switchover from one source of a nutrient to another. It appears that when the bacteria change nutrient they need to undergo some internal changes that enable them to take up the new resource. The hypothesis is that this switch takes time. What is nice about this explanation is that it can be extended to include the lag phase at the beginning of the experiment. The idea is that at the start of an experiment the cells need to switch from no nutrient to a particular nutrient. This explanation could work if we assume that cells are not normally ready to take up any nutrient source unless they are exposed to it. Only in the presence of a particular nutrient will they switch on the necessary systems to metabolise and take up what they find. The lag is the time it takes to power up the bacterial machinery necessary to take up and use nutrients.

There is one complication though with this explanation. In the first experiments in figure 3 where the bacteria are grown on glucose there is no lag phase. One may first try to explain this away by assuming that the glucose metabolism is always turned on. This explanation does not hold, unfortunately, because in the "one pot-two nutrients" experiments where the cells initially grow on glucose there is a lag phase in the beginning. This indicates that the absence of a lag-phase is nothing inherent to glucose. There is a slight variant of this explanation, that is more consistent with the data. Instead of assuming that cells are always ready to take up glucose, maybe the glucose machinery had already been turned on in the first experiment, but not in the second. This could be because the cells, were exposed to a glucose medium before the start of the experiments. It turned out that this was indeed an explanation that was consistent with all the experimental findings.

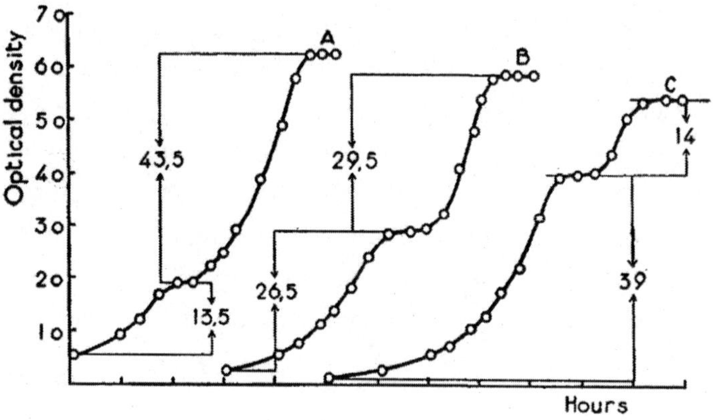

Figure 3.2: Diauxic growth as observed by Jacques Monod (Monod, 1949, figure 9). The three curves correspond to experiments with 50/150, 100/100 and 150/50 μg of glucose/sorbitol each. (Permission obtained)

Altogether, Monod's experiments seem to indicate that bacteria have specialised uptake mechanisms for particular types of nutrient. When the cells are capable to use glucose then they are not able to take up xylose as well. Indeed, they will shun xylose as long as there is glucose in the system. If glucose runs out then they can switch from one nutrient source to the other. However, this switch takes time, which manifests itself in the lag-phase. What precisely is going on internally to the cells, Monod did not know.

From his simple growth experiments Monod could learn a little bit about bacteria and how they function internally. His systematic experimental enquiries were certainly progress compared to a simple stamp collecting bacteriology. When he built a laboratory system and systematically varied conditions, Monod went beyond mere observation and did true experimentation in the spirit of Galilei. Monod could now give a very detailed description of how the bacteria grow, what they do when offered several sources of nutrient and that they somehow need to re-adjust themselves internally when preparing to use a particular type of nutrient. However, Monod knew nearly nothing about the details of the cellular processes that bring about the observed behaviour. What is the machinery that needs to be switched on to take up nutrient? How does the cell "know" there is nutrient in the environment? How is this "knowledge" translated into switching on the machinery?

A mechanic, if confronted with a machine that he does not understand, would open the machine and check how it works. In the case of cars or lawnmowers and any other type of machine this may work, provided one is skilled enough in dealing with mechanics. With cells the same strategy may still be successful, but is certainly much harder technically. Monod could not just undo a few screws and simply inspect the internal mechanisms of *E.coli*. He faced two main difficulties: For one, the components of interest are very small. That in itself made it impossible for Monod to find them, extract them and examine their properties. Then, what made it even harder, typically biological components do not work outside the context of the cell, or at least they are not always amenable to a functional analysis outside the cell. Mechanics do not face these problems. A suspension can be tested even if it is not installed in a car. An isolated gear-box shows all the properties of a gear-box whether or not it is built into the car or sits on the shelf.

For this reason Monod had to come up with a novel way of analysing systems. New methods that allowed him to identify and describe components. Monod had to take a very different approach to what a mechanic would have done. Instead of disassembling the system and investigating the components one at a time, Monod used a method of controlled manipulation of the system. The essence of what would become his methodology was to make controlled changes to the organism and then see how this impacts on their behaviour. The key to doing this successfully was to choose the right kinds of changes that would provide maximal insight into the mechanism. If the particular manipulations were well chosen, then the reaction of the system to the manipulations could give substantial insights.

An important aspect of this experimental method is to form hypotheses about conceivable mechanisms for phenomena he wishes to explain. Hypotheses are conjectured explanations or educated guesses about how something might possibly work. Coming up with these hypotheses requires a good bit of imagination and the ability to integrate this with already secured knowledge. The art of experimental science is to generate the right kinds of hypotheses, those that are general enough to provide deep insights, but at the same time they must be sufficiently concrete to constrain the design of new experimental tests to corroborate or indeed refute them. Hypotheses function as mental crutches and almost visualise what is going on at the level of the cell. Hypotheses are indispensable elements when discovering the inner workings of nature.

Monod knew that he had to focus on a very specific phenomenon if he wanted to find out about the agency of the cell. He decided to focus on the so-called lactose system. Just like arabinose or glucose, lactose is another type of sugar. And just as in the case of these other two sugars, *E.coli* has to undergo some internal changes in order to be able to take up and use lactose for growth. Specifically, Monod decided to focus on this research question: What precisely is it that allows the cell to take up one nutrient, but not others? How does the cell know which nutrients are in the environment and how does it regulate the transition between different conditions?

In many ways the starting point for Monod was the discovery of a specific molecule, called β-galactosidase. He noticed that this molecule is always present in cells that grow on lactose. So, his initial hypothesis was that β-galactosidase may be somehow connected to the machinery that is necessary for lactose uptake and metabolism. It was unclear to Monod what precisely β-galactosidase was doing, although it seemed reasonable to assume that it was somehow involved in processing the lactose and converting it into energy for the cell. So, this was one component identified.

Monod was also able to identify the uptake system for lactose. He called it galactoside permease. Using radioactively labelled molecules Monod found that the cell only takes up lactose when this permease is present. These initial findings made it quite clear to Monod that both β-galactosidase and the permease are essential for the bacteria to show the exponential growth dynamics when fed on lactose. He also knew that these two components were only present in the bacteria when the bacteria were grown in lactose, but absent when grown on different sugars. So, it seemed that β-galactosidase and the permease system were important parts of the machinery that needed to be switched on in order to be able to process lactose. This observation was a powerful starting point for Monod and allowed him to condense the rather general question about the agency of living systems to a concrete research problem: How can the cell "know" that lactose is in its environment? What are the internal mechanisms that allow *E.coli* to turn the production of β-galactosidase and the permease on and off? Do these mechanisms generalise to other systems, say to the regulation of glucose and arabinose uptake?

In his investigations Monod found a powerful assistant in evolution. He observed that sometimes in his growth experiments bacteria started to change. As he was growing his cell lines, he would notice that from time to time he ended up with strains that had lost certain abilities that other bacteria had. The existence of these variant bacteria is a consequence of the action of natural selection i.e. random changes in the DNA of the organisms that led to variant behaviours. Fast growing species like *E.coli* with reproduction times of 20 minutes can evolve in front of our eyes and develop new features, new properties. Monod could take advantage of this. Whenever a suitable type of mutation came along, he grabbed the chance and conserved the strain. These mutants supplied by evolution turned out to be crucial tools in his quest to understand what is going on inside the cell.

One of the variants Monod had discovered was the noGal mutant. This mutant had lost the ability to produce functioning β-galactosidase. Therefore, noGal bacteria that are fed on lactose cannot grow anymore because they have no way to metabolise the nutrient they are offered, although they still grow on other nutrient sources. By labelling the lactose radio-actively, he could show that these cells still take up lactose, but they are not doing anything with it. The nutrient just accumulates in the cell without being used. Another mutant he found was noPerm. This mutant had the ability to produce galactosidase, so could in principle metabolise lactose. However, it was unable to produce the permease, and hence no lactose reached inside cell. Finally, Monod also had a third mutant, the noOff mutant, which could produce both the permease system and glactosidase. In fact it only differed from the wild type in that it produced galactosidase constantly, also in the absence of lactose. The noOff mutant lacked the switch to turn off the lactose machinery.

There was another crucial tool that Monod had at his disposal. Bacteria such as *E.coli* have the ability to exchange genetic material. Normally reproducing bacteria simply divide which means that the offspring is a clone of the parent, maybe with some minor variations caused by mutations. This is different from, for example, humans, where the offspring receives genetic material from two individuals, father and mother, and is not genetically identical with either. The ability to exchange and mix genetic material can, however, be evolutionarily advantageous sometimes. Bacteria have evolved mechanisms to transfer individual genes from one cell to another cell that is not its offspring. There are several mechanisms by which they do this. One of them is a process called conjugation, whereby two bacterial cells come together. One of them then injects a number of genes into the other. After conjugation the recipient—female—cell has an additional set of genes, namely those given to it by the donor—male—cell. These genes function normally alongside endogenous genes of the recipient. One can think of these genes as coding for very specific functional modules that can be freely passed between bacteria almost like software packages that can be passed between computers.

Monod was able to utilise this natural tendency of bacteria to exchange genes. He used this to add functionality to individual bacteria. The key was again to do this in such a way that he could learn as much as possible about the invisible internal mechanisms of the cell. Comparing the properties of cells that had only one set of genes with those that had both allowed him to make inferences about the mechanisms by which genes work.

In one of the first experiments he performed, Monod combined a noGal mutant with a normal non-mutant bacterium, or *wild-type* cell as they are commonly referred to among biologists. In these experiments he exploited conjugation so that the noGal strains injected their genes into those of the wild-types. So, technically, the wild-types now gained new functionality. Only that Monod expected that this combination would have no effect whatsoever. The piece of DNA containing the noGal mutant could not produce β-galactosidase, so probably did not add functionality at all. The wild-type therefore should be unaffected by the new set of genes and continue to do what it would do in their absence. As long as the bacteria are not exposed to lactose nothing is produced. Once lactose is introduced, then the wild-type genome is used to produce β-galactosidase, whereas the new components received from the noGal mutant cannot produce anything, so do nothing. Altogether, no difference is expected. Similarly, if a noPerm mutant is combined with a wild-type, the combined system behaves just like the wild-type. Monod did the experiments, observed no effect and was satisfied that his method of combing genes seemed to work.

With these preliminaries in place, Monod started to direct his attention to the actual question: How does the cell regulate the expression of the proteins that are necessary to metabolise lactose? How does it switch on the machinery? How does it switch it off again? He wanted to know what precisely it is that leads to the sudden production of the permeases and β-galactosidase. To address this question, Monod thought that the noOff mutant could be very useful. This variant cell had apparently lost the ability to regulate the induction of the lactose system, so it could be instructive to learn more about it and how it differed from the wild-type strains. Unlike wild-type bacteria, its machinery is always on, ready to use lactose. Monod wondered now what would happen if he combined the two mutants, a noOff mutant as the donor and the noGal mutant as the recipient cell. Technically, the procedure is the same as in the case of the combination of a wild-type and a noGal, only that now Monod expected to see something. Before the gene transfer the female (recipient) cell cannot grow on lactose. On the other hand, the male genes are incapable of turning off the lactose system. What happens when they are combined?

Monod did this experiment, let us call it EXP1, and got a clear result. After transfer of the male genes into the female, the new cells that now combined the noOff and the noGal mutants (although on different DNA strands) behaved qualitatively like the wild-type. This is a startling discovery. When on their own, the two mutants are either completely dysfunctional or unable to regulate the production of lactose, but in combination they can compensate for one another's shortcomings and regain both the ability to produce β-galactosidase but most of all also the ability *not* to express it when it is not needed. If nothing else, this shows that the regulation of the machinery and the machinery itself can be separated from one another without losing functionality.

To find out more, Monod performed a variant of this experiment, let us call it EXP2. He used a male cell that had both the noOff and the noGal mutation. The male cell is dysfunctional and cannot grow on lactose. As a female cell he used a wild-type cell. Clearly, the female cell on its own can perfectly function and turn the lactose metabolism on and off. When he combined the two cells, he found that the cells that had both genomes behaved like the wild-type, being fully functional and able to regulate the expression of β-galactosidase. Now, at first this seems like a rather dull outcome. Nothing happened when the foreign genes were injected into the female. However, the experiment allowed Monod now to make an important inference about how the lactose machinery is regulated. The *prima facia* conclusion is that the noOff mutant is recessive: When this mutant is combined with another mutant that is *not* noOff then it has no effect on the cell and the combined cells retain the ability to control the expression of β-galactosidase. Beyond the surface interpretation, the experiments revealed two crucial facts about how β-galactosidase is regulated. Firstly, the regulation does not appear to be local. Even though the noOff mutant and the noGal mutant in the first experiment were on different genes, they could still influence one another. This is clear from EXP1 where none of the individual strands of DNA are by themselves able to function, but in conjunction they do.

It is worthwhile dwelling on this a bit more. In EXP1, one of the cells had the ability to control the expression of the lactose machinery, and the other cells had the machinery itself. Both are somehow coded for on the genome, but the design of the experiment guaranteed that their genes resided on different strands. The question is now whether or not one can deduce useful knowledge from that. Certainly Monod thought that this is possible. His preliminary conclusion was that the regulation of genes somehow must involve global signals in the cell. Presumably some sort of chemical message that in some way or another can communicate the presence of the relevant genes across the cell.

One can think of this in terms of a car. Think of the noOFF mutant in EXP1 as a car without a steering wheel and the noGal mutant as a car without an engine. What Monod did in EXP1 was to take the steering wheel of the one car that has no engine to control the other car that has no steering wheel. Clearly, that would not work because a steering wheel and an engine need to be connected to the same car. In Monod's words, there must be a local connection. Now imagine instead a remote controlled car. Then the noOff mutant is a car without remote control and the noGal mutant is a car without an engine. Clearly, it is possible to take the remote control of the noGal car to steer the noOff car. Remote controls can work over a distance. There is no need for the remote control to be physically connected to the same unit as the engine.

EXP1 demonstrated to Monod that whatever controls the expression of β-galactosidase this is some sort of remote control. The controlling unit does not need to be physically connected to the controlled unit. If you have a remote control you also need some sort of signal that transmits the control commands. Monod immediately hypothesised that it is a protein that is released into the cell volume (the cytoplasm) and acts anywhere in the cell. Yet, based on the experiments he had performed so far, it was impossible for him to say anything specific about the control signals.

In fact, he could not even be completely certain about the remote control idea. It is fully possible to come up with many other explanations that would be compatible with the experiments. In this sense, nothing has been proven by Monod's results. Still, he provisionally accepted the idea because it was a simple explanation that also fitted well with what else was known about cells and their functionality. Biologists know that they have to come to conclusions in the absence of ultimate proofs. It is rarely possible to exclude all loopholes of a particular interpretation of experimental results. If one finds an explanation that fits the available data, the explanation is simple and integrates with what is known otherwise, then this can form the basis for further work. Ultimately, though, there is no certainty.

Having accepted the remote control ideal, the next question Monod asked was about the nature of the regulator. This question was addressed by EXP2: The regulator must be a repressor, at least in its combined effect. To formulate it differently: The difference between the wild-type cells and the noOff is that the latter lacks a repressive remote controlling signal, that is molecules that suppress the expression of β-galactosidase and the permease system.

This conclusion emerges directly from EXP2 and the observation that the noOff mutation is recessive. The wild-type has the repressive signal that is lacking in the noOff mutant. Therefore, when the wild-type cell obtains the genes of the noOff mutants, then there is no effect, because the former already has the signal that *prevents* the expression of β-galactosidase, the signal that noOff does not have. Now clearly, if the regulator were an activator, then the same experiment would not work. The noOff genes, once combined, would start to produce their activator in the new cell and turn on β-galactosidase production. The combined cell would behave exactly like the noOff. This means that injecting the noOff genes into a wild-type cell would permanently activate this cell, which is not the case. Hence, Monod learned that the so far unknown regulator of β-galactosidase must be a repressing signal.

The repressor signal produced by the wild-type cells can be compared to jammers, the type that repressive regimes sometimes use to prevent their people from listening to radio stations from neighbouring countries. These jammers work in essence by emitting a signal that interferes with the radio waves arriving from the foreign country. The wild-type cell can be thought of as a prospective listener, let us call her Ms. A, who waits for broadcasts from abroad which she can then pass on to her fellow suppressed citizens. Normally the evil regime makes sure that these jammers are always switched on, preventing our would-be revolutionary from receiving any news. So, in order to be able to listen to the foreign news Ms. A and her fellow resistance fighters need to bribe the radio-engineer responsible for the jammer, to produce a temporary "failure" of the device during the broadcast of the main news from abroad. These bribes have the same effect as lactose in that it represses the repressor/jammer. Now Ms. A can listen to the foreign radio station and pass on unbiased news, as long as the engineer receives his bribe. There is of course an alternative possibility to enable her to listen. The local underground resistance movement could destroy or sabotage the jammer. If they succeed in doing this, then the bribes/lactose are no longer required and Ms. A can listen to the news and spread her message in the local community. This is the analogue to the noOff mutant.

The noGal equivalent is when Ms. A and her activities attract the attention of the secret police and she gets arrested. This means that there will be nobody to listen to the news and spread them to the community. The local people will not be able to get any unbiased news irrespective of whether the jammer is on or not.

To understand Monod's experiments in this analogy, we have to assume another secret listener, Ms. B. She is also keen to receive unbiased news from abroad and also relays whatever she hears to her local community, exposing the lies of the regime and replacing them with the unbiased truthful news from the neighbouring country. Of course, Ms. B is equally prevented from doing so by a jammer, a different one though from the one that prevents Ms. A. If we assume that these two brave women are geographically separated by large distances then any jammer sabotage at Ms. A would not affect Ms. B's ability or inability to listen to the news. Analogously, whatever genes are turned on in one cell cannot affect the biochemistry of another cell. However, the situation would be very different if Ms. A and Ms. B lived close to one another. Say, they lived close enough to feed the same local environment with news and also close enough to be affected by each other's jammers. This would be the analogue of the genes of one cell being injected into the other cell.

Monod's EXP1 considered the case where Ms. A was arrested (the noGal mutation) and the jammer of Ms. B was off (the noOff mutation). As long as Ms. A and Ms. B are geographically separated, the local environment of Ms. B would receive news, but not the local environment of Ms. A. This is the case of the cells being separated. The analogous effect of injecting the genes of the noOff mutant into those of the noGal mutant is that Ms. A and Ms. B feed the same community and are affected by both jammers. In this case the local environments of Ms. A and Ms. B, which are now the same, do not receive any news. Ms. A is in jail and cannot report anyway. Ms. B is prevented by the remaining jammer ("Ms. A's" jammer). If she wants to listen to foreign news, the engineer at Ms. A's jammer needs to be bribed with lots of lactose. The fact that Ms. B's jammer is destroyed does not help at all.

This is the picture Monod could work out: Biologically, it seems that the cell is repressing the expression of β-galactosidase and the permease, unless they are needed, rather than activating them when they are needed. This is rather peculiar. The cell seems to be set up such that it "naturally" is ready to express everything that is necessary to use lactose, but then it also has something else that prevents it from actually doing it. This is a bit like a car that is always on full power but with a break engaged as well to prevent it from rolling. In order for the car to go somewhere the break must be released. It may seem more natural to have β-galactosidase and the permease system being off unless there is an activator.

Why is the system organised in this way rather than in a more natural way? Are there any benefits to de-repression over the conceptually simpler activation? Legitimate questions to ask, but not the ones Monod focused on. Instead of speculating about underlying reasons for particulars of organismic design, he was interested in more practical issues. Not content with merely understanding that the lactose system is repressed under normal circumstances, he wanted to find out about the precise mechanisms of this repression. The experiments he had done up this point are compatible with at least two models. For one the jamming model. Since Monod now knew that the repressor is a global signal, a molecule that is present in the entire cell volume, it is conceivable that the repressor somehow interferes with β-galactosidase molecules and the permease system, just like the jammers of evil dictators interfere directly with the radio signals from neighbouring countries. Alternatively, it could be that the repressor is not a jammer, not a signal that interferes with another signal. Instead, it could be that the repressor directly interferes with the ability of the cell to produce β-galactosidase and the permease system. So, in the analogy of the evil dictatorship, this would be the case where the evil government, instead of jamming the incoming signals, directly sabotages the emission of the signals from abroad. They could for example threaten the radio engineers in the free country with abduction and death and torture should they allow the signal to be switched on (this is admittedly not a very realistic scenario).

Note that the experiments Monod had conducted so far could not distinguish between the jamming scenario and the alternative—let us call it the sabotaging scenario. In principle it should be possible to distinguish between them. For the sabotaging case we need to imagine that there are two evil agents operating in the free country, both keeping up the pressure on the radio engineer. Even if one of them gets uncovered and arrested by the good policemen of the free country, the other evil agent would still be able to operate and effectively prevent transmission of the radio signal. In this scenario the mutations would have slightly different analogues. For example, instead of bribing the jammer-engineer, the resistance group needs to bribe the evil agent abroad, so that she stops threatening the foreign engineer. The reader can easily convince herself that this story and the one with the jammer are indistinguishable in terms of their experimental effect. The only difference between them is precisely who needs to be bribed, or biologically speaking, what precisely lactose is doing in the system.

There are a number of arguments why the sabotaging-scenario, is a more plausible alternative, at least when it comes to biology. An important consideration is the energy household of cells. Biologically, jamming is expensive. Most of the time there is no lactose in the system but the cell still needs to produce jamming molecules and β-galactosidase, only for the former to deactivate the latter. This is quite an energetic burden and feels like a huge waste. Based on this consideration the sabotaging scenario seems much more plausible. If there is no lactose in the environment, the cell has no need to expend the energy needed to produce or maintain the whole machinery it needs in order to metabolise it. Hence, it is best to avoid suppressing its production rather than just its action.

Sabotaging seems a better alternative, but mere plausibility arguments are not enough to convince a scientist like Monod. He wanted to generate some experimental proof. New experiments had to be conceived. With the experimental tools available at the time this was very difficult. Like Galilei before him, he needed to go to the limit of what was doable. Monod had to push beyond the state of the art. There was no way for him to directly observe what is going on in the cell. So, again, like Galilei, he had to design experiments that gave him the necessary information indirectly.

Monod hoped that he would be able to get a better insight into the mechanisms of regulation if he could get a better feel for the sequence of events during repression and de-repression. He was particularly interested in a time-resolved picture, something that could give him a feeling for the duration of the individual steps. Such a timed sequence of events that take place in the cell as it is exposed to various conditions could go a long way towards constraining possibilities. In the early 1960s when Monod was active biotechnology started to get to a point where it became feasible to monitor the chemical composition of the cell under well-defined conditions.

Monod was determined to take advantage of the technological possibilities of his time and to measure precisely what is going on in the cell as the machinery to take up and metabolise lactose is turned on. He decided to measure the time it takes to turn on the β-galactosidase gene. One way to do this is to simply put a colony into lactose and measure the amount of β-galactosidase over time. The problem with this is that there are potentially many things that happen between the cells being exposed and the switching on of the genes. For example, lactose must first be taken up. It is unknown how long this takes. Since Monod was only interested in the time it takes to switch on a gene, he had to somehow factor out all these potentially conflating processes that distorted his measurements. Somehow, he needed to define a time zero when the gene switch is flicked, so to speak.

He found a way to do this using his standard trick: The transfer of genetic elements between different strains. His setup was like so: Monod constructed a line of cells that combined the noOff and the noGal mutations. These cells would continuously express β-galactosidase (because they are noOff), but they do not actually express it because they are noGal. In the above analogy, these cells correspond to the case where Ms. A is in jail, but there is free transmission of the foreign radio (either because the jammer is broken down or because the secret agent had been uncovered, depending on which scenario one accepts). This construction was a good starting point for Monod. Even though they lacked the ability to express β-galactosidase, they still expressed the permease. So, they accumulated lactose internally. This line of bacteria was precisely what Monod wanted. The cells are now ready to switch on the β-galactosidase genes the moment they get them. The permease is expressed, lactose is accumulated and everything is ready to go, except that the cells did not have the crucial genes for β-galactosidase—yet.

Next, Monod mixed these cells with another line of wild-type donor cells. The idea was that these donor cells could now transfer the genes necessary to express β-galactosidase to the mutants, which has the effect of switching on β-galactosidase expression. This defined the time 0. All he needed to do then was to monitor the amount of β-galactosidase in the cells—which he could do—to get an accurate idea of how long it took to turn on the genes while disregarding any conflating factors. There was one loose end to tie up. In the presence of lactose, the wild type donor cells would themselves produce β-galactosidase. This is conflating and undesirable, and defeats the purpose of the experimental design. The problem was that he would not be able to distinguish between the β-galactosidase produced by the double mutants and the donor cells. A small tweak to the design could circumvent this problem. Monod introduced an antibiotic into the broth in which the cells live and he ensured, by careful selection of the cell lines, that only the donor cells, the wild-type, were sensitive to this antibiotic. It would not kill them, but it would prevent them from expressing any genes. So, even though they had the ability to express β-galactosidase they did not do this because of the hostile environment. Now he had a clean experimental design. Both strains of cells, by themselves, could not express any genes, they could not produce β-galactosidase. Yet, the donor cells could transfer their genes to the recipients, and with both genomes the recipients could then start to produce β-galactosidase straight away. No delay from de-repression. Since the donor cells were sensitive to the antibiotic, Monod could also be certain that any β-galactosidase he did measure would be from the double mutant cells that had received the additional genes.

With his experimental tricks he managed to initialise the cells in a way that suited him and he was then in a position to observe how the amount of β-galactosidase produced by the cell changed over time as conditions changed. Based on his previous experiments, Monod expected that the recipient cells, once they received the genes from the wild-type, behaved rather like the wild-type cells. That is, in the presence of lactose they produced β-galactosidase, otherwise they did not. This is indeed what he found to be the case—at least in the long run. Rather surprisingly,

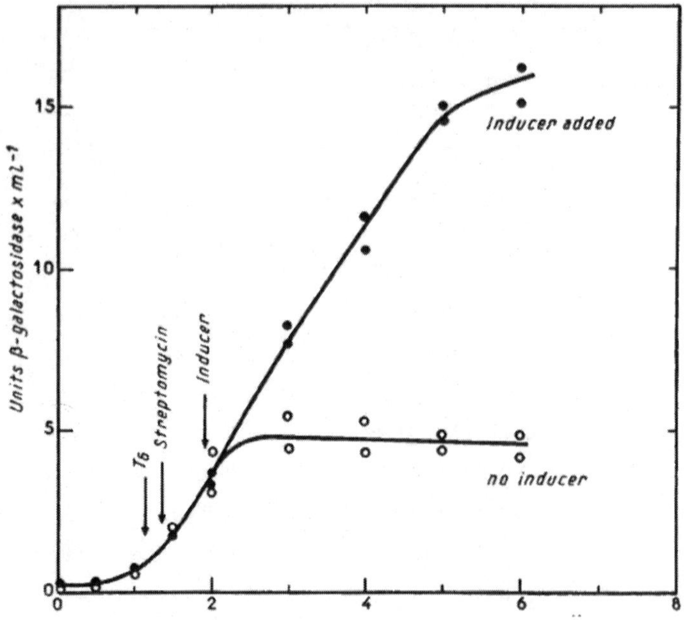

Figure 3.3: The data from the original experiment performed by Monod (Monod et al., 1959). The horizontal axis shows the time that has passed since starting the experiment, the vertical axis the amount of β-galactosidase produced. The graph shows clearly that the production of β-galactosidase abruptly stops after about an hour unless lactose is added to the system. (Permission obtained)

the cells behaved differently in the short run. The moment they started to receive the genes from the donors, cells started to produce β-galactosidase even if there was no lactose around in the system. This phase did not last forever. After an hour or so, he could still get the cells to produce β-galactosidase but only if he added lactose to the system. So, at this point the cells acted qualitatively similar to wild-type cells in that they would only produce β-galactosidase when the lactose was present.

There are at least three insights Monod could gain from this experiment. Firstly, β-galactosidase production can start immediately without delay, if the conditions are right. As soon as the necessary gene was in the cell, the bacteria start to make the protein within a few minutes. This indicates that the process of gene expression is fast to switch on. Secondly, it seems that the repressor only becomes active with a time delay. For the first hour since adding the necessary genes, the cells with and without lactose as an inducer behaved identically. This seems to show that the cell does not accumulate enough of the repressor to allow efficient repression of β-galactosidase. Finally, and this is perhaps the most interesting part, the production of β-galactosidase is interrupted rapidly once the repression becomes effective. In the graph in figure 3.3 this is indicated by the curve that indicates the amount of β-galactosidase in the bacteria without inducer. It seems that the hypothetical repressor of β-galactosidase needs to accumulate to a critical concentration before it can do anything at all. This is why the cells with and without lactose in figure 3.3 are identical for a long time. However, once the inducer has reached this threshold, it does its job straight away and quite effectively and the curve without inducer flattens out rapidly. The flat part of the curve indicates complete arrest of β-galactosidase production. The crucial question was now what Monod could learn from this about the repression of β-galactosidase.

The observation that the production of β-galactosidase is

interrupted nearly instantaneously once the repressor becomes effective is difficult to reconcile with the jamming model where the repressor interacts with β-galactosidase. The measurements show that it takes a long time to switch off β-galactosidase production, however, once the process is started it does not take a long time before β-galactosidase production is completely arrested. Once the no-induction curve starts to deviate from the induction curve, it flattens off very quickly. A jammer would behave differently. It would slowly inactivate β-galactosidase as it accumulates in the cell. The two curves would start to diverge from one another soon, long before β-galactosidase is completely suppressed.

The curve Monod actually observed suggests that the repressor slowly accumulates in the cell. Once it has reached a certain threshold concentration, it then becomes active and represses the production of β-galactosidase, unless some inducer is in the cell. So, it seemed to Monod that the sabotage model better accounts for the activation of the lactose metabolism in *E.coli*.

One aspect that complicated the interpretation of the data was the lack of information about the precise number of steps that are involved in gene expression. Monod had to assume that it was a multi-step process. The problem was that he did not know what the individual steps are, nor how many of them there are. This added considerable uncertainty to his model. For example, it was certainly conceivable that the repressor acted on the DNA directly to prevent β-galactosidase from being expressed in the first place. Alternatively, it could also be the case that the repressor interacted with some intermediate cell products between DNA and the protein. The experiments Monod had conducted up to this point were not able to decide between these possibilities and it was not clear how this could be addressed experimentally. Yet, in order to be able to understand how the cell works, it was essential to get some understanding of this.

Then Monod made a crucial discovery. Among his cells he found a new mutant, the noOp mutant. At first this mutant looked similar to the noOff mutant. The noOp mutant produces β-galactosidase even in the absence of lactose, although as Monod found not as much as the noOff mutant under the same conditions. So, at a first glance it appeared to be identical to a noOff mutant. A closer look, however revealed that it is very different.

To work out the details Monod used another mutant, the galP mutant. This mutant is normal in all respects, only that it produces, instead of β-galactosidase, another form of the enzyme that can be chemically distinguished from the original form. He used this galP mutant to check in detail how the noOp mutation affected the cell. Specifically, he wanted to know precisely how the presence or absence of mutations on a particular string of DNA impacts on the expression of genes both on the same DNA strand and on others. This comes back to the question about the locality of gene interactions. He had clearly demonstrated before that the recessive effect of the noOff mutant is due to non-local action in the sense that it is associated with repressor molecules that apparently pervades the cell. The question now was whether the noOp mutant also repressed non-locally. With this β-galactosidase variant, he could decide this.

Monod combined a genome with the noOp mutation and a galP mutation with a wild-type genome. He now had one cell, with two genomes, two strings of gene-beads so to speak. One with the galP mutation and the noOp mutation, and one that is the normal wild-type. He wanted to find out whether the noOp mutation would have any impact on the wild-type chromosome, and if so, what this impact would be. What Monod found was both surprising and revealing. It turned out that these cells always expressed the variant of β-galactosidase but they only expressed the proper form when lactose was added to the broth. This seemed to indicate that the noOp only acted locally on the genes on the same DNA strand, but it did not influence the expression of genes located somewhere else. To be sure, Monod then turned the experiment around and used a noOp mutant located on the same DNA as the wild-type for β-galactosidase and combined it with a genome that had a galP mutation but *not* a noOp mutation. He found now that the resulting cells always expressed the proper β-galactosidase, but only expressed the variant when lactose was present. Had Monod performed the same gene-shuffle experiments with the noOff mutant he would have found that the order would have been irrelevant. Whether or not the noOff mutation is associated with the galP or the wild-type β-galactosidase would have made no difference whatsoever.

This clearly demonstrates that the noOp mutant corresponds to a mutation of the genome that is directly implicated in the regulation of the expression of β-galactosidase, but only locally. The noOp mutation only affects the genes on the same string of gene-beads, but has apparently no effect on the expression pattern of genes on different strings. The implication for Monod was that there are (at least) two types of regulations, or two components to regulation in the lactose system. On the one hand there was the mystical repressor that seems to be ever present in the cell. It is coded for by a genetic structure that is somehow compromised in the noOff variants. It clearly behaves like a global signal that pervades the cytoplasm and acts non-locally, as established previously. Then there is the regulatory element corresponding to the second mutation, the noOp mutation, that acts only locally. This suggests that it does not code for an enzyme that then subsequently moves through the cytoplasm, but somehow changes local conditions in the proximity of the genes to enable or disable expression of proteins. But how? What is it doing? How come the repressor action is both local and non-local?

The experiments themselves do not give direct evidence about the underlying mechanisms. However, Monod did not need this. He could simply assume some scenarios, create hypotheses. As one does in these circumstances, he first considered the simplest possibility. To begin with he assumed that (i) the repressor in the lactose system somehow represses by doing something to the DNA directly. So, the idea is that the repressor—in an as of yet unknown way—interacts with the gene-bead that codes for β-galactosidase and thus prevents it from being expressed. (ii) The noOp mutation is then a modification of the local structure of the DNA that, somehow, prevented the repressor from repressing. It could be, for example, that the repressor needs to bind to the DNA to prevent the expression of β-galactosidase; in the presence of the modification it may be unable to do that and β-galactosidase is continuously expressed.

This hypothesis, although far from certain, would account for the experimental evidence and can explain why there is both a local and a global aspect to repression. However, it also has a very specific implication: The repressor of the lactose system acts at the level of the gene. In the light of the experimental evidence Monod had, this seemed to be a plausible assumption but nothing has been proven yet. At least not beyond all doubt. Having arrived at this point, Monod now felt some urgency to get more experimental data to provide uncontroversial proof that β-galactosidase is indeed repressed at the level of the gene. So he devised an ingenious experiment (Monod et al., 1960), the *gene removal experiment*. The hope was that this experiment would show conclusively how the repressor represses.

It did so, although it did so very indirectly. It is important to appreciate this point in order to understand precisely what Monod was doing. Rather than probing how the repressor worked, the gene-removal experiment added further constraints to the interpretation of available experimental data, including the noOp experiments. These extra constraints then allowed some deep insights into the mechanics of gene regulation. The *prima facia* conclusion Monod would be able to draw from the gene removal experiment was that the life time of intermediate gene products must be short. This sounds irrelevant to the question at hand, but is indeed very relevant. Control of the expression of β-galactosidase at the level of the gene only made sense if the intermediate products exist for a short time. Monod comments on this as follows:

The most immediate and also perhaps the most striking of these implications is that the structural message must be carried by a very short-ived intermediate both rapidly formed and rapidly destroyed during the process of information transfer. This is required by the kinetics of induction. As we have seen, the addition of inducer, or the removal of co-repressor, provokes the synthesis of enzyme at maximum rate within a matter of a few minutes, while the removal of inducer, or the addition of co-repressor interrupts the synthesis within an equally short time. Such kinetics are incompatible with the assumption that the repressor-operator interaction controls the rate of synthesis of stable enzyme-forming templates[...]. Therefore, if the genetic operator model is valid, one should expect the kinetics of structural gene expression to be the same as the kinetics of induction: injection of a "new" gene into an otherwise competent cell should result in virtually immediate synthesis of the corresponding protein at maximum rate; while removal of the gene should be attended by concomitant cessation of synthesis.(Jacob and Monod, 1960, p. 461)

The basic idea of the gene removal experiment is simple. First take a gene that is actively expressing an enzyme (in Monod's case β-galactosidase). Then remove the gene while it is expressing and check what effect this has on the expression rate. There are two possible outcomes. Either, the expression of the enzyme stops immediately following the removal (this is what happened). In this case one can be certain that any intermediate products between gene and the finished proteins would be short-lived or non-existent. Or, there is a delay between the removal of the gene and the cessation of protein production, which would indicate that there are some longer lasting intermediates in the chain from gene to protein. This second possibility would mean that regulation at the level of the gene could not function rapidly in the following sense: Even if the gene is removed, then the long-lived intermediates would still be around for some considerable time and continue to be transformed into protein. However, previously Monod had observed that β-galactosidase production stopped more or less immediately once the repressor had built-up beyond a threshold concentration. Therefore, should Monod observe long-lived intermediates in his gene removal experiment, then this would indicate that the hypothesis of the repressor directly interacting with the gene-beads is not compatible with the available data which clearly indicated that genes can be switched off rapidly. In this case, Monod would have had to conclude that the repressor somehow interacts with a long-lived intermediate product rather than directly with the DNA.

The gene removal experiment addressed the question about the precise regulation mechanism only indirectly, but nonetheless conclusively. Theoretically the experimental strategy was elegant. The real problem was its realisation. How does one remove genes from a live cell without destroying it? To do this Monod had to mobilise the entire experimental armoury of the time and use his theoretical ingenuity to design a credible experiment. In retrospect the solution he eventually found seems both elegant and simple, but it took intellectual titans to come up with it.

Actually removing a gene in the sense of cutting out a piece of the DNA of a live cell without killing it was well beyond the technical abilities of Monod at the time. Rather than actually removing the gene in a controlled way, he found an arrangement where he could slowly decay the gene in question using radioactive radiation. As the source for his radioactive radiation he used a specific isotope of phosphorus with a half-life of 14 days. The half-life is crucial, as it will turn out. If it is too long, then the experiment would go on for too long, which is not desirable. If on the other hand the radio-active decay was too rapid, then it would have been too difficult to control the experiment. The radioactive radiation would inflict significant damage at the preparation stage of the experiment. He did not want that. He needed a source that allowed him to grow cells for hours without being affected, or at least without being significantly affected by the radio-active decay. In this respect, two weeks is a good time.

A suitable half-life is great, but only the beginning. If he wanted to do anything with the radio-active material he needed to get it into the genome. This is what made phosphorus so suitable for his purposes. Phosphorus can be very easily inserted into the DNA. Once inserted, it would stay with the gene-beads for a very long time. When it eventually decays, then the DNA is destroyed but only locally. At least with a certain probability, this destroys the information contained on the DNA.

At this point Monod knew how to destroy DNA in the living cell. There was one remaining problem though. He needed to target the damage of the radio-active decay specifically to the β-galactosidase gene-beads. The experiment would be pointless if the radiation destroyed genetic information indifferently. At first this seemed like an unsurmountable problem, because he did not have a technique to limit the insertion of phosphorus to specific sites only. However, as it turned out he did not need to. Instead Monod could use his standard tool of gene exchange between strains that has served him so well in the elucidation of genetic control. The idea is straightforward, but also shows his experimental acumen. Monod took one strain of bacteria and inserted a great amount of phosphorus into the DNA without worrying about where in particular the radio-active labels ended up. The only thing he was concerned about was to cover as much DNA as possible. This procedure gave him a cell whose DNA was densely packed with radio-active phosphorus labels. Here comes the crucial part. He now used the mating techniques to transfer sections containing the genes relevant for the lactose system into a strain whose DNA is otherwise unlabelled. The recipient cells now had an unlabelled DNA, except for the additional genes for the lactose system that were radio-actively labelled.

For this experiment to work Monod had to rely on the law of large numbers. He could not be sure that any individual piece of DNA transferred from one strain to the other is labelled with the radio-active phosphorus. He only knew that the DNA of the donor cells contained was densely packed with labels which means that each stretch of DNA itself had a high probability of being labelled. This in turn means that a high fraction of the DNA pieces that were transferred had the phosphorus tag. To make sure he could see any effect, he had to rely on the fact that in his broth there was a large number of cells that mated and hence he had also a high number of cells that transferred the labelled DNA. At this point it becomes also clear why a suitable half-life of the phosphorus was so important. The mating can only be successful if the cells themselves, both the donor and the recipient cells are functional. Given that it was necessary to heavily tag the donor cells, a rapidly decaying label would have destroyed them during the preparation phase of the experiment, making the experiment itself impossible to perform.

With the preliminaries settled Monod could now concentrate on the design of the experiment that would help him address his scientific question. He prepared two different cell lines. The recipient cells contained the noGal mutation. They did not produce any β-galactosidase by themselves. The second donor cells were wild-type except that they were heavily tagged with radio-active phosphorus and they were also sensitive to a specific antibiotic. As a result of this antibiotic the donor cells would not produce any β-galactosidase by themselves even in the presence of lactose. The idea of the antibiotic sensitivity is the same as in the noOp experiments above. Monod simply wanted to be sure that any β-galactosidase produced in his system was produced by the recipient cells, and not the donor cells which were of no particular interest. Next Monod mixed the donor and recipient cells in the same broth containing the antibiotic and let the exchange of genes happen, as had already become routine for him. He then waited a bit and checked that the recipient cells, now complemented with the wild-type genes, started to produce β-galactosidase, as they should. And they did. So far the experiment did not go much beyond what Monod had done previously.

All these preparations and testing took some time, but not enough for the phosphorus to do considerable damage to cells. So, by the time the cells started to express β-galactosidase Monod had merely convinced himself that they still worked despite the heavy manipulations and the radio-active labels. The gene removal proper had not started yet, although the main work had been done at this stage.

To continue the experiment, Monod froze the cells. This arrested the cell processes but not the radio-active decay. While frozen the phosphorus would slowly disintegrate and with it the DNA that it labelled. This would destroy the gene-beads coding for β-galactosidase, while the cells were frozen. After two weeks Monod thawed the cells again expecting that a considerable proportion of the cells would have their genes coding for lactose destroyed, or "removed." Apart from that, the cells once thawed were in exactly the same state as two weeks earlier. They would simply have awakened from a two week long hibernation. As soon as they had unfrozen, they would continue to do whatever they had begun before being frozen. Only that there was no more DNA for β-galactosidase. The gene was removed.

This was now the moment of truth. If there is indeed

only a short lived intermediate between the DNA and the protein, then one would expect that after thawing the β-galactosidase production is immediately arrested. Translated into the specific experimental setup this would mean that the rate of β-galactosidase production is much lower after thawing than before freezing. The reason for the reduced production, rather than the arrested production is that not all strands of DNA were labelled, and not all labelled ones may have been destroyed by decay. In any case, Monod did observe an immediate reduction of β-galactosidase expression in the thawed cells. And he took this as final evidence that the intermediate products are short lived.

The Monodian way of doing biology

Much has been learned about gene regulation since the days of Monod. However, the conceptual picture of regulation that Monod developed is unchanged to this day. The lactose system served Monod well as an example of gene regulation. The picture that emerged from it turned out to generalise quite well to other genes. One of the important features of the lactose system that Monod found was that it consists of several genes, all of which are regulated by the same signal. He called such a group of genes an *operon*. In the lactose operon that Monod studied one repressor prevents the expression of both the permease and β-galactosidase. Those two genes (and as it turned out later there is a third one as well) are always expressed together, or indeed not expressed together.

The model we have today is that the expression of genes, and hence the production of individual protein species, is modulated by other proteins in the cell. These regulatory proteins are called *transcription factors* (TF). TFs can regulate in two basic ways. They can enhance or stimulate the expression of the protein they regulate, or they can suppress it. In the case of the lactose system Monod identified a repressor. To understand how it works a little bit of background is needed.

Whenever the genes of an operon are to be expressed, the gene sequence must be read. The cell has specialised proteins, molecular reading heads that fulfil this task. The reading head scans along the DNA and converts the genetic information into protein (in a multi-step process). When the reading heads are not reading they are freely suspended in the cell liquid. In order for the gene sequence to be read correctly, the reading heads need to know where the gene starts and where it ends. It would be no good to read only the second half of a gene. For this purpose the cell has some signals placed on the DNA, the *promotor sequence* that tells the reading heads that they can bind and start to read. These signal sequences are very much like the guide-lights on run-ways at airports that indicate to incoming planes where they should land.

The lactose repressor manages to prevent expression of the genes in the operon in a very simple way: It binds to the promotor sequence, and covers the guide lights. The reading heads no longer attach to the crucial part of the DNA and will not express the genes. When lactose comes into the cell, then it displaces the repressor from its binding site. This makes the guide lights visible again and ensures that the genes for the permease system and β-galactosidase are expressed. In Monod's noOff mutant the gene-bead coding for this repressor was mutated. As a consequence the cell could no longer produce the protein necessary to repress β-galactosidase. Even in the absence of lactose in the cell, it no longer occupied the promotor site. The result was that β-galactosidase and the permease system were permanently expressed.

The noOp mutation was a mutation that prevented the repressor to cover the guide-lights. So, even if there is a repressor around in the cell it did not do anything. This is also the reason why the noOp mutation acted locally on the specific strand of DNA only. The noOp mutation only covered the specific guide lights on the particular strand of DNA.

This regulation mechanisms has one obvious problem. The expression of β-galactosidase (necessary to metabolise lactose) and the permease (necessary to take up lactose from the environment) is suppressed unless there is lactose in the cell. This means that if there is no lactose in the cell then the cell cannot take up any lactose in the first place. So, in a sense to detect lactose in the environment you need lactose in the cell. Then again, if there is no lactose in the environment, there cannot be any in the cell either. To get around this an unavoidable fact of bacterial life comes to the rescue. In bacterial cells, nothing is certain. Even if the β-galactosidase gene-beads are perfectly repressed, from time to time the odd protein will still be expressed. Normally the repressor hides the guide-lights, but sometimes it falls off. When this happens, then normally another repressor molecule will take its position and quickly hide the guide-lights again. Sometimes, however, in this short time window of opportunity when the repressor is off, the cell manages to quickly read the gene and produce a little bit of β-galactosidase and a few molecules of the permease. Even if there is not a single lactose molecule in the cell, sometimes things just happen and a bit of permease is produced. Most of the time the cell has the ability to take up a bit of lactose, even if the relevant genes are repressed. This little bit of lactose is sufficient, however, to de-repress the production of β-galactosidase and permease. When this happens, then the floodgates are open. More lactose can enter the cell, de-repressing the system even more until β-galactosidase and the permease are produced at maximal capacity.

Monod's approach

The influence of Monod's work can hardly be underestimated. The lactose operon is still the best studied operon in bacteria. His hypothesis driven approach to experiments will be familiar to any contemporary microbiologist. The sophistication of methods has changed since; knowledge has increased, but the basic approach to knowledge generation is still the same. What is most important though, Monod was among the pioneers of a totally new way to do biology. Instead of stamp collecting, the Monodian approach to biology is more Galileian. To understand biological systems as biochemical and biophysical systems whose behaviours and properties can be understood in terms of interactions of particles. Unlike the stamp collectors before him (or indeed after him) Monod did not investigate organisms in the wild, in the kinds of environment where one would naturally find them. He did not explore guts to find *E.coli.* Instead he confined his bacteria into the laboratory. He created well defined, controlled and homogenous conditions. Most of all he created environments whose conditions were known to him. Under these known conditions he then investigated what the cells were doing, how they function internally.

Monod could only do this because he had the ability to make very particular manipulations of specific aspects of the cell. This could be the inclusion of radio-active labels into the genome, particular mating properties of his cells and specific sensitivities to antibiotics. The experiments required him to have full control over the amount of lactose that is contained in the assay, to be able to add antibiotics as required, and the right amount. He also had to be certain that there were no other nutrients in the assay that interfered with the bacteria. In effect, Monod needed to shield his system from environmental fluctuation. He had to create an environment that was defined, conditions that he knew and that he could control at will. He had to create a laboratory system.

The experimental systems Monod used were idealisations, similar but not equal to the idealisations of Galilei and other physicists. In this sense, Monod's systems were quite far removed from reality and constrained the organisms under study to conditions that they will not normally encounter during their natural life. The danger is that these artificial environments introduce behaviours that are not relevant in natural environments. Biologist routinely study micro-organisms under conditions of plenty and under conditions of scarcity. This is a bit as though an alien species interested in finding out about us would study humans by confining groups of people into Gulag like conditions or into golden palaces with all luxuries supplied. Clearly both of those conditions would lead to behaviours that are of little relevance to the typical behaviour we show in ordinary life situations. The same applies to bacteria. Natural colonies rarely grow exponentially; indeed, they probably never find conditions where they can grow exponentially for extended periods of time on a defined pure medium. They will practically never grow in a single species colony. Realistic environments are nearly always multi-species, where *E.coli* will have to compete with other types of bacteria.

Monod found himself in the same situation as Galilei. In order to do anything, he had to idealise. It would have been impossible for Monod to perform his experiments in realistic environments. For one it would be very difficult to perform experiments on mammalian guts one of the environments where *E.coli* naturally occur. The first problem is to access the environment in a convenient way. This is a methodological issue that is in itself insurmountable. Even if he solved that problem, he could not be certain that measurement results are relevant and about *E.coli*, rather than any of the countless other species that populate the gut. Even if all this was not a problem, there would still be no exponential growth in these natural environments. Instead of being able to perform experiments within a few hours, every sample population would have to be observed over very long periods of time simply because natural environments do not support the rapid growth that is possible under lab conditions. Non-exponential growth is slow and drags out the period over which the system needs to be observed. On top of that, another problem is that there would be no defined medium.

Monod could not have been certain that the bacteria actually take up lactose at any one time. If there are other nutrients in the environment (and invariably this will be the case), particularly nutrients that the cell prefers, then the cell would not take up lactose, even if it was there. Remember the case of pizza and humous. In an undefined environment it is very difficult to be confident about what precisely the bacteria are doing. From a research perspective unknown conditions are sub-optimal.

But let us assume that Monod somehow would have managed to overcome all these obstacles. He would then have faced another problem. How would it be possible to measure quantities such as, say, the amount of β-galactosidase after a certain amount of time has passed? To be precise, measuring the quantity would be no problem, but it would be difficult to interpret it because Monod could never be sure how many bacteria there are and how much of the β-galactosidase he measured is produced by the *E.coli* rather than any of the other countless species that can be found in natural environments. The optical density can only be measured when the cells are suspended in a well stirred and transparent medium. Natural environments are neither well stirred nor transparent. Furthermore, they are multi-species, so any measurement would wildly over-estimate cell numbers belonging to the species of interest.

For all these reasons Monod would have been unable to conduct his experiments in a real environment. Idealisation was essential. He had to shield his organisms from the environment. This means placing them into an environment that he could control. Taking the bacteria out of the context in which they would normally be found and placing them into an artificial environment, a bottle with a defined amount of nutrient and a defined temperature. By doing this he severed the connection between the bacteria and their natural environments.

Idealisation is not only about controlling the environment itself, but also about controlling the organism. In his experiments

Monod used artificially made variants of wild-type species. All the mutants Monod used were in a specific way insensitive or overly sensitive to particular conditions. So, for example, in one of the experiments the donor cells were very sensitive to an antibiotic, whereas the recipient cells were not. The noOff mutant is insensitive to the presence and absence of lactose in the environment. The noOp mutant is also insensitive to lactose, but in a different way. In Monod's experiments the characteristics of these mutants were crucial to tease apart conflating effects. A particularly good example of the latter is his experiment to measure the time required for genes to be expressed. It would not have been possible to simply record the β-galactosidase content over time in a wild-type and infer the expression time from the graph. The reason is that in the unmodified bacteria β-galactosidase production is repressed and one has to assume that de-repression takes time. It is not known how much time this takes. If Monod had simply recorded the amount of β-galactosidase over time, then this would have conflated the time required to take up and de-repress the gene and the actual time to express the gene after de-repression. In this particular experiment he was only interested in the latter and had to find a way to direct his experimental enquiry specifically onto this aspect, and nothing else. His laboratory system, for this experiment, had to shield the phenomena he was interested in from all other processes that may go on in the environment and within the cell. For these experiments, the relevant system was no longer the organisms as a whole, or the colony, but the particular process of expression of the gene coding for β-galactosidase excluding its regulation.

When he used his combination of donor cells that are sensitive to antibiotics and recipient cells with the noGal and noOff mutation, Monod shielded the expression of β-galactosidase from the other aspects of β-galactosidase. These modifications of the wild-type cells had played the same methodological role as the polished balls and ridges or inclined planes that Galilei used to shield his experimental system from the effects of friction. Or indeed the layers of liquid nitrogen and helium in the Müller and Peters experiments to prevent thermal expansion of the resonator.

Theoretical models

Like the physicists before him Monod generated knowledge from idealised experimental systems, what we call laboratory systems. The spirit of Galilei has infused Monod. A part of Monod's drive to construct laboratory systems was pragmatic. Monod would not have been able to answer the questions he asked if he had not used laboratory systems. He was not content with pure observation and description as it is done in stamp collecting biology. He wanted to find a mechanistic explanation of life. He wanted to demystify the agency of living systems and explain it in terms of physical interactions. Monod and his successors in biology made large contributions to our understanding of the biological systems in terms of their underlying components. Still, the biology that resulted had only aspirations to be like physics. To this day this ambition is not achieved.

This manifests itself in the lack of predictive theory. Galilei, but also Müller and Peters had very particular ideas as to what the results of their experiments were going to be. This expectation was rooted in a theoretical understanding of their systems. Based on a mathematical description of the free fall Galilei had very specific expectations about the behaviours he would observe in his experiments. He also knew what types of idealisations his laboratory system had to reflect so that the experiments would be a fair test of the theory. Galilei knew that he had a number of particular problems to overcome, that he needed to remove friction and that he needed to measure time accurately enough. The idealisation Galilei had to find was driven by the assumptions of the mathematical theory. The same is true in the case of

Müller and Peters.

Monod did not have a theory that he could use as a chart to steer his experiments. At least at the beginning of his research, he could not have any expectations about how bacteria work internally. To some extent his experiments were a flight in the dark. As he went along he started to form hypotheses about what was going on, what drove the phenomena he saw, what kinds of underlying events could generate them. At no point, however, did he have a general and unified theory similar to Newton's laws or Einstein's theory of relativity. The laws and theories in physics make statements that are independent of the particular instantiations. They only work because they are general. For Einstein's theory of relativity it does not matter whether the traveller is a bacterium, somebody's twin brother or an inanimate clock. The same effect of time dilation holds irrespective of what is travelling at a high speed.

Biology is different. Monod could only make hypotheses about his particular system. His experiments could only uncover what *E.coli* is doing, and more specifically how the lactose operon in *E.coli* works. If he wanted to say something about another organism, he would have to check that whatever he learned in *E.coli* was still true. Indeed, he did not even have any guarantees that the mechanisms he discovered in the lactose systems would generalise to other operons in *E.coli*. Strictly for every regulated gene the regulation mechanism needs to be researched separately, one by one.

Monod's program was to understand the agency of living systems in terms of underlying physical interactions. In a restricted sense he succeeded. His experiments did uncover the molecular basis for some of the behaviours *E.coli* shows. Still, he did not reduce biology to physics. Even now, many decades after Monod, there are no unified and general models of biological systems.

Interactions

Monod's approach to biology has been seminal in biosciences. Since his original work in the 1950s and 60s great advances have been made. Biologists have now vastly refined ways to probe and observe cellular processes. For example, the resolution of imaging techniques is so good these days that researchers have recently succeeded, for the first time, to visualise a single repressor in the lactose system (Elf. Lie, and Xie, 2007). This and similarly powerful techniques are now creating the possibilities to measure every aspect of the cell, to understand and observe cellular processes in much detail and even to model them mathematically. Does that mean that we are about to understand the agency of living systems in terms of physical interactions. Or at least that biologists are on the way to getting there?

Biology is heading towards an ever more detailed understanding of biological processes. Assuming this development continues into the future, eventually research will uncover the full physical basis of biological systems. Researchers will have a complete list of parts of particular biological systems. Even now new components are discovered nearly on a daily basis. New molecules and interactions are added to the data-bases that hold biological information. Sequencing the entire DNA of an organism is now a matter of routine. It is very likely that the day will come when we know it all. At first we will probably know all about *E.coli* and then soon about other organisms as well. When this happens the Galilei-Monod program of biology is completed. Will we then have a complete understanding of life?

No, what we will have is a long list of components and no idea what to do with them. To generate useful knowledge biologists would still need to know how things connect together, which gene interacts with what, which protein does what, how different genes work together to collectively create the agency that we observe.

Understanding the components is not the same as understanding what the system is doing. Just think of flat-packed furniture. These days, beds and wardrobes are often sold as components, for the consumer to assemble them at home. When you buy one of those, you do not get the bed or the table, but planks, screws, connectors, poles and so on. The parts contain everything needed for the wardrobe or the bed, but you cannot put your clothes into it or sleep on it before the individual components are assembled. These flat-packs are the furniture equivalent of a complete list of *E.coli* components.

The advantages of selling furniture in parts are clear: Efficient production processes, cheap and easy transport, low storage costs and so on. All this means better furniture for less money for the consumer. The disadvantages are clear as well. Assembling flat-packs is not always trivial. Putting the parts together in the right way can be challenging for the practically less gifted.

The consumer is not without help. For one, there are the assembly instructions that are meant to be a fool-proof step-by-step guide to the finished furniture. Needless to say they hardly ever are. Then, the parts themselves have some fool-proofing built into them. There is a limited number of ways in which planks and boards can be put together. The holes for screws and the shapes of connectors often constrain the way things can be assembled. If the owner tries to put together two boards that are not meant to be connected in that way, then most of the time she will not get very far. Still despite all this fool-proofing and help, assembling flat-packs can be very difficult.

Now imagine a super-sized flat pack. Not just an individual piece of furniture, but the contents of your entire house as a single huge flat-pack. Not only the wardrobes and beds, but also the shelves and racks in your kitchen and bathroom as well. All delivered as components, individual cables, screws, and springs, fans and brackets for you to assemble. Imagine now that all these components come without the instruction manual, all supplied in a single box without labels. What a nightmare this would be to assemble.

Arguably contemporary science, particularly biology is in precisely this situation. Data-bases hold terabytes of information about organisms. Knowledge that has been painstakingly acquired during thousands of research lives. Yet, what does it all mean? How do the parts slot together? How can we assemble the bits and pieces into a coherent picture? What does all the detailed flat-pack knowledge tell us about real life? We do not know yet.

It appears that the Galilei-Monod paradigm of biology has an internal contradiction built into it. On the one hand, it is a very efficient way of producing reliable knowledge. The history of science has shown this. On the other hand, the outcomes can be self-defeating. The knowledge creation can become so efficient that we are inundated with information that we cannot process any longer. At the same time, away from the highly idealised systems that physics looks at and that naturally generalise, the Galilei-Monodian way of doing science tends to a sprawl of knowledge, the creation of larger and larger flat-packs. These flat-packs are in some sense complete sets of information, but without any context. By themselves, the individual pieces of knowledge that are created do not help us to understand organisms better. Knowledge of all the genes and proteins and how they interact probably gives very little information about the behaviour of the organism. The living breathing animal is reduced to a list of components. Its properties get lost within all the flat-pack detail. What is more, if a Galilei-Monodian approach comes to define what science is or at least what scientific knowledge is then this directly discourages anybody from finding out about the real thing. After all this would only be stamp collecting, a clearly inferior form of research. The more we know about the physical basis of the agency of life, the less we know about life itself. The Galilei-Monod science, if consequently applied, steers into a black hole of ignorance.

Even if we see this, it is hard to stop the drive that is intrinsic to the Galilei-Monodian paradigm. If we stop researching before reaching the ultimate explanation, then a strong sense of dissatisfaction would remain. The Galilei-Monod approach has an inbuilt momentum towards the information abyss. As long as we describe phenomena in high level terms, there will always remain another "how-question" to be asked: How did the repressor find the binding site? How can β-galactosidase metabolise lactose? How does the permease allow the cell to take up lactose? If we take the Galilei-Monod paradigm seriously, then we can only be happy once the biological system has been reduced to some fundamental physical theory. There is no premature exit, there are no stop signs. If we refuse to drill down further, if we declare satisfaction with the knowledge we have then the program has ultimately failed. We would implicitly admit that, say, proteins and their properties are unexplainable. This would be a rather frustrating conclusion. On the other hand, we can agree to go along and embark on the next research program, attempt to explore even more detail. We will then, as we indeed have, learn more about the fundamental properties of biological systems. Unfortunately, we then also create even more difficult flat-packs. Instead of every plank and board of our scientific furniture being delivered as a whole, we are now given flat-packs with saw-dust and glue and varnish to build and shape our own boards. Needless to say, there is still no assembly instruction. The deeper we drill down the levels of nature, the closer we come to achieving the goal of the Galilei-Monod program in biology, the harder it will get to re-assemble the components and pieces into a coherent whole, or to make sense of them. The research program loses sights of its aim. The organism and its agency disappear as more and more detail of its physical function is uncovered.

There is another problem. Central to the Galileian approach to science is the idea of the laboratory system. Laboratory systems are at the same time necessary pre-conditions for efficient science, but they also narrow down the scope of the outcomes of this science. In physics laboratory systems are essential in order to re-create the idealised conditions of the theory that these experiments set out to test. In the case of Monod it was slightly different. There the function of the laboratory system is still to screen environmental influences, but not so as to create a theoretically conceived idealised system, but in order to create experimental conditions that can be interpreted. The laboratory systems are not driven by the demands of the theory, but they are methodological necessities.

This is now also where potential problems start. There is absolutely no guarantee in biology that it always makes sense to study laboratory systems. Imagine that an experiment can only be done when certain manipulations of systems are performed or when certain simplifications are made. These manipulations could lead to conditions that are completely irrelevant for the question one wants to address. In biology this could be the case when systems are isolated from their environments as it is so often necessary for laboratory systems. Organisms and their behaviours depend crucially on the interactions with their surroundings. Unlike laboratory systems, real environments are never constant, but are constantly changing. Think, for example, of the great treks of African wild-life through the savannah, in search for food. Thousands of Gnus and Zebras face perils and predators in their migrations across the African planes. It would be impossible to understand these migrations by studying a single Gnu in a controlled environment, for example an enclosure in a zoo. The cage is an idealisation that makes the behaviour (great trek) unobservable. The behaviour of the animals in the wild is a result of the particular conditions of the environment in which the Gnus live. Zoo-Gnus do not migrate. Not only because they are in an enclosure, but also because they have no need to migrate. There are no migrating cues for the Gnu in the zoo.

There is now a contradiction. If we want to study migrations using a Galilei-Monod approach the system must be idealised, yet the idealisations sometimes remove the phenomenon we want to understand in the first place. This is true for the Gnu, but it is also true for many other things in biology, even many aspects of bacterial behaviour. The lactose system can be studied by isolating the bacteria from their environments, but even these simple organisms have social lives that are very difficult to understand in a laboratory system. Real environments are nothing like the homogenous nutrient saturated land of Cockaigne that bacterial laboratory strains so often find. Indeed, it is not even clear that the behaviours of a laboratory-rat *E.coli* are the same as those of their brethren in the wild. Many of these strains have specialised to laboratory life and have acquired mutations that differentiate them in crucial ways from clinical strains. Indeed, there may not even be a single clinical strain. Real environments are highly differentiated and colonised by variants of the species that are adapted to these specific conditions. Understanding interactions of this sort is very difficult, if not impossible using a Galilei-Monod approach, but often requires biologists to go back to do stamp collecting again. This does not mean that the Galilei-Monod approach is bad or should not be used. It is not. It can lead to great and important understanding. However, it is not the only way to generate knowledge in biology.

Chapter 4

Complexity

Mehr gabst du, Wunderfrau,
als ich zu wahren weiss:
nicht zürne, wenn dein Lehren
mich unbelehret liess!
Ein Wissen doch wahr' ich wohl:
dass mir Brünnhilde lebt;
eine Lehre lernt' ich leicht:
Brünnhildes zu gedenken!
Götterdämmerung, Prolog

Billiards

There are several types of the game of billiards. The most common one is American pool, where each player has 8 balls, either stripes or solids. The idea of the game is that the player whose turn it is uses her cue to shoot the balls into one of the pockets along the cushions of the table. What makes the game interesting and challenging is that the players are not allowed to directly strike "their" balls. They can only do so indirectly via a shared white ball, the so-called cue-ball. The cue-ball has to hit the players' own balls at just the right angle to expedite them into one of the pockets. This rule allows all kinds of tactical complications that ultimately make the game interesting. A common strategy is for one of the players to leave the shared white ball in a position that makes it hard for the opponent to do anything useful. For example, she may not have a direct line from the white ball to any of her own balls, which reduces her options for the next shots.

Another effect of the rule is that it makes the gameplay much more challenging. Imagine all players can directly pocket their balls, one by one by directly striking them with the cue. This would be too easy to be interesting. Controlling the solids and stripes indirectly is much more difficult. With the intermediate it is a wholly different ballgame. A little uncertainty here, a small error of judgment there, a tremble, a momentary lapse of concentration is enough to take the white ball off the intended course. The error is amplified when it hits the target ball.

Despite this chicane of the intermediate ball, for experienced players American pool can become trivial to play. Good players eventually learn to clear the table in one go. They pocket the first ball, then the second, then the third and so on until the last ball. For those masters pool is no longer the challenge it is for the novice player.

There are several options to make the game more interesting. Either, one could make the table bigger, the balls smaller, the rules more challenging or all of the above. Snooker is perhaps the most difficult variant of billiards and has all of these features. The difference between a Snooker table and a normal American pool table is impressive, certainly for those who have never played this game before. The distances to the pockets are much longer and consequently more precision is required. If this was not enough, the balls are also smaller, so getting the angles right is much more difficult. The pockets themselves are smaller too and shaped in a less forgiving way. An inaccurate shot reflects the ball away from the pocket, rather than into the pocket as in many American pool games. Snooker also has rules that are designed to minimise the role of chance. It is much more unlikely that the player will benefit from unintended shots. Lucky shots are rare, more often they are unlucky shots and lead to penalty.

Another way to make the game harder is to introduce more intermediaries. A variant of billiards that is popular amongst the intellectuals in the formerly smoke filled rooms of the cafés of central Europe is Carambolage. It is played on a table with no holes and only three balls. A white ball, a dotted ball and a red ball. Each player has either the white ball or the dotted ball. The aim of the game is to hit the opponent's ball and the white ball with one's own. In the most challenging form of the game the rule is that the player's ball has to hit the cushion at least three times before it hits the second ball. So for example, if the player owns the white ball then, in order to get her point, she could hit cushion, cushion dotted ball cushion red. Or she could hit the cushion three times first and the red and dotted ball. Needless to say, this is a very hard game.

Where does this difficulty of gameplay come from? Clearly, it has something to do with the interaction of parts, that is balls in the case of billiard. A single ball on the table is easy to control. Three balls are much harder. Interacting with three balls and three cushions requires a master player. If even more interactions are introduced, then there quickly comes the point where it is impossible to control anything. Assume we changed the rules of pool so as to require two intermediate balls, rather than just one. Say, one first has to hit the white, then an (assumed) yellow ball and only then one can hit stripes or solids. This difficult version of pool would be a very frustrating game to play. Now imagine that we increase the number of interactions further. Rather than just two intermediates, let us require three or four or even more than that. As it turns out, the point comes when the game becomes unplayable. This new game of billiards is not only hard to play, but in fact impossible to play. Even an ideal player, with no trembling or shaking and a perfect understanding of the angles required would ultimately fail at doing anything but lucky shots. The smallest inaccuracy at the first ball amplifies with every intermediate to the effect that after just a few of those it is *in principle* impossible to predict where the last ball is heading. Once the level of accuracy has to be better than the size of an atom, the billiard game becomes random.

What is true in billiards is also true in science. Nobody knows this better than physicists. They love their simple theories, laws of nature with neat mathematical expressions to go with. If you take a closer look at those simple theories, then you will quickly notice one thing. There are no interactions in these theories, at least not very many of them. Take Galilei's square law. In a strict sense the fall of an object in a gravitational field is the interaction between two entities, the falling body and Earth which exerts the gravitational field. Indeed, the distinction between a falling body and an attracting body is artificial, because technically the two bodies fall towards one another. Given the difference in size between the two, it is however reasonable to ignore this fine point and approximate the big body (Earth) as stationary. The effect is that one interaction is removed and the equation gets much simpler without any detectable error.

On a cosmic scale the same trick does not work any longer. Two very large masses (for example two planets in close proximity) mutually attract one another. They literally fall towards one another. It requires some skill to work out the general solution for this, but ultimately for a trained physicist this is not a big deal. Most second year students of physics can do this.

Let us add another "cushion." What about three interacting bodies? Now the problem has completely transformed. It is still possible to calculate the trajectories of three bodies, but it is no longer possible to write down a mathematical formula that gives the solution. With three bodies or more the trajectories need to be simulated on computers. Add even more planets and the problem gets harder and harder and any hope to get a nice formula to describe how bodies move is long gone.

Physicists have now a problem. On the one hand they like simple formulas. On the other hand they have to face the realities of life. There are no formulas for problems that involve more than two bodies. What can they do? The simplest strategy, and this is the one that is usually followed, is to simply avoid those problems. Deny that they exist and you are fine. Physics is the most fundamental science, the science of mathematical beauty. If something cannot be expressed using maths it has no place in the church of Galilei. Leave it out. Ignore it, deny its existence. This is how physicists came to believe that the world is built on simple principles.

To be fair, many physicists do not insist on mathematical formulas alone to predict the behaviour of their system. Many applied physicists are perfectly happy to calculate solutions numerically when exact formulas cannot be obtained. There is, however, one thing that gets lost when one does simulations. Generality. In a sense, simulation is stamp collecting the physics way. Any solution one obtains is valid for one particular system only. It cannot be generalised to other problems. No new laws can be generated from simulations. It is perhaps practically important, but it is low prestige work in physics. Therefore, once there are significant interactions involved Galileian physics hits its limits.

Wait a moment, a physicist may now object. This is not true. Some of the most fundamental of physical laws are very much about interactions. These theories are foundational and they are formulated in terms of neat simple formulas. Think of thermodynamics which describes how gasses behave, how they expand when they are heated up, what their pressure is and so on. Its insights are so general that they even find applications in fields as diverse as biology, computer science, chemical engineering, chemistry and astronomy. This is a truly fundamental theory after the heart of the physicist. At the same time it is all about interaction. The fundamental theorems of thermodynamics are derived by considering how gas molecules interact with one another and their environment. The air you breathe is made of millions of millions of molecules, all of them colliding with one another many times per seconds. It is like a giant, 3 dimensional billiard game played on a planetary scale.

If interactions would make it really so hard to predict, so the objection goes, then how is it possible that there are simple and general theories about the behaviour of gasses? How can it be that one of the best and most accurate theories in physics is about this seemingly hypercomplex planetary billiard with its billions of colliding parts, while at the same time we struggle to come up with a simple equation to predict the motion of three planets?

The answer is as follows: Thermodynamics is feasible precisely because it consists of so many particles and interactions. The number of molecules in a gas is so large that we no longer need to care about what the individuals are doing. Thermodynamics is a science about the typical behaviour of large aggregates of particles. It is still the case that interactions are difficult to predict and once there are enough of them they are impossible to predict. It is not feasible to calculate the trajectory of a particular molecule in a gas, where it goes, when it collides with other molecules and so on. Nobody can do that. The point is, however: Nobody wants to do this. This is not what thermodynamics is all about. Prediction is impossible, that is prediction at this fine-grained level where gasses indeed look like giant billiard games. Thermodynamics is not about solving the giant billiard problem. It is not about charting out the interactions of the individual molecules and following them one by one. Instead, it is about understanding the large scale features of the game, how many balls there are in a specific region of space *on average*, how often the balls collide with the cushion *on average*, how fast the balls will fly *on average*. Nobody cares about the position of Fred the oxygen molecule at 9am on Monday. This is completely irrelevant. The world of thermodynamics is a collective world where everybody is equal to the point that they are devoid of individuality. One molecule is like any other. All that counts is their collective average behaviour. In this society of clones, interactions are the great equalisers. When two molecules collide then they exchange their speed and direction of flight. They keep the statistical diversity of the system in a way that is predictable.

Thermodynamics is about the large-scale features of gasses. Pressure and temperature are in essence averages that emerge when millions and millions of particles collide many million times a second. These averages are only predictable because there are so many molecules and so many interactions. The thermodynamics of a real Snooker table would be much less accurate. It is precisely the planetary scale that saves the day for the thermodynamical billiard game. Things are predictable when there are few, very few interactions, or maybe none at all and when there is an extremely large number of them. The problematic area is the middle ground, systems with many interactions, too many to remain tractable, but not enough to be able to disregard them or just look at the average.

On fish and termites

The problem for the physicist is that most real systems are in this difficult middle ground. They are complex this sense. The Galileian paradigm does not work. Take as an example a shoal of fish. It consists of hundreds of individual fish that somehow interact with one another. Hundreds is clearly too many to be tractable for physicists, but not enough to create the nice average behaviour of millions of molecules. The fish do not behave like a gas. They are more akin super-organisms when they swim in near perfect synchrony apparently driven by a single mind. They have no leader, no general that guides the troops. Still, the shoal remains in strict formation. When they encounter obstacles then they nimbly divide, just to re-unite once the obstacle is mastered. Predators that attack the swarm are avoided with a clever defensive strategy that leaves the attacker confused and likely unsuccessful. How can these fish keep a coherent swarm? How do they determine where to go? How do they dazzle the predator?

Another example are termite mounds. These insect dwellings

are highly functional structures, built without an architect, without a central planner or supervisor. They are ventilated structures that auto-regulate both temperature and air quality within. They have specialised rooms, including the queen's chamber in the inner-most part of the mound, farming areas where the termites grow fungi and thick outer defence walls that protect them against a variety of enemies. Our intuition is that structures of this sort do not come about by accident. They need to be conceived, engineered, carefully planned and then built by an organised workforce. There is no evidence though that termites are capable of such planning, nor do we have any reason to believe that they have the basic engineering skills required to conceive such a structure. How come then that they can still build their mounds?

To put this into perspective. Imagine you want to build a house and hire some workers for it. You are not allowed to tell your workers where you would like to have your house built, nor will you tell them what the house should look like, what rooms it will have nor where the rooms are. Indeed, you are not even allowed to tell the workers that you are building a house and that they need to collaborate in doing so. It seems unlikely that such a project will be successful and will ever result in a house. Yet, this is precisely how termites build their mounds. As it will become clear below, the key to their feat is interaction. Individual termites are like simple programmed machines that can react to their environment and other termites, and they do so in just the right way so that a termite mound results.

Insects like termites would not have the cognitive abilities to engage in the type of central planning and complex organisations that are required to build the skyscrapers of Pu Dong or Manhattan. Nevertheless their structures can be up to 300 times their size (assuming a termite is about 7mm long), which is about the same ratio as the Taipei 101 skyscraper to an average human (assumed to be 180cm). Keep in mind that the termites are successful in building their own skyscrapers by just following their instincts. They do not have the highly organised workforces, nor the machinery, nor the infrastructure that is necessary for us to build our own. Termites are too simple to have desires or wishes or plans. They are not clever enough to plan and execute a centrally planned engineering venture. Still, they can build skyscrapers. The termites' way of building is very different from ours. They build in a de-centralised way. Rather than having a single chief-engineer who directs the army of workers, the termites direct themselves. They follow simple rules that evolution has implanted in them.

It is difficult to imagine how a de-centralised building site can work, a site where nobody is in charge and nobody makes decisions. Consider this: Whenever a bricklayer starts to build a new wall, first of all she has to decide on its location. This will involve some groundwork where she prepares the surface on which the bricks will be placed. In order for the task to be successful she must have an overall idea of the finished wall, where it should be and in which direction it will go. Most of all, she must have an idea that she is building a wall, rather than a kiln or nothing at all. The plan for the finished product will inform every choice she makes during the building process. Where to lay the first brick and where the second brick goes in relation to this and the third and so on. Every single step in this process is motivated by the intention to build a particular wall with particular features. If the builder had no plan she would not know what to do. Termites do not have this intention. Still, termites build impressive structures. How can they do that when they have no plan?

Unsurprisingly they start off by laying bricks randomly. No plan, no blueprint, no engineer, only one choice. This sounds like a bad strategy. One would think that depositing random pieces of material only results in rubble...and it does. However, even in randomly deposited heaps of rubble, there is some internal structure that can be exploited by the termites.

The principle is very simple, but rather different from what human engineers would do. Before depositing a brick, termites impregnate the brick with a pheromone. Pheromones are chemical signals that are quite common in the animal world. One can think of them as scents. The main point is that pheromone deposits can be sensed by other termites and affect their behaviour. Termites are attracted to the pheromone signals of other termites and are more likely to to deposit their pheromone impregnated bricks in a place that already has a strong scent. As it turns out, this attraction is key for the success of the termites to build complex structures without any central planning or control.

The pioneers in the field of the social behaviour of insects are the French scientists Eric Bonabeau and Guy Theraulaz. They were among the first ones to decode the behaviour of termites and ants and study how simple interactions between the individual animals can lead to them building elaborate structures even in the absence of a plan or a chief engineer. They found that the crucial starting point for any termite dwelling is the rubble field. Bonabeau and Theraulaz suspected that a well-known principle from physics—symmetry breaking—could explain how the random rubble can be turned into a structured building. The important point to bear in mind is that random does not mean equal across the area. When bricks are deposited randomly, then, just by chance, there will be areas with more bricks and areas with fewer bricks. Since all bricks are impregnated, this also means that there are areas that have a stronger scent than others. The laws of chance demand this. Uneven distributions of this sort can be observed in any rubble field. There is no magic about that.

From close-up the initial rubble field looks like an undulating landscape of heaps of bricks. There will be lots of bricks in one place and no bricks in other places. All this is the result of the random placement of building material. Remember now that every single brick that the termites deposit is impregnated by pheromone. So, for the termites the undulating landscape of rubble is also an undulating landscape of scent. The termites are programmed to deposit the next brick in one of the areas that have a strong pheromon scent. As a result, this small heap gets a bit higher and has a somewhat stronger scent. The next termite will then be even more likely to deliver its cargo at this location as well. The overall effect is that naturally arising fluctuations of brick assemblies are amplified. The more bricks there are in an area, the more likely it is that even more bricks will be deposited. The winner takes it all. Wait long enough and the termites have decided, collectively, on where to build the structure and they will build it. All this without a project manager, engineer or indeed a plan.

Bonabeau and Theraulaz could show in computer models how this simple preference rule leads to the construction of pillars and walls. There is more to the story. As the pillars and walls are erected, the environment changes. The structures have an influence on how the pheromone scents are distributed in the nest, and change in this way also the way in which termites are attracted to it and the way in which they add building material to the emerging structure. These changes of the environment can modify the structure that is erected from say a pillar to a wall. While the rules that the termites follow are always the same, the environmental cues change and hence the shape of what they are building. Bonabeau and Theraulaz give an example. The queen, the egg-laying centre of every colony, herself emits a scent that entices termites to build a protective wall around her. Again, they follow exactly the same rules as they did to build the pillar. Only, now the environment is different and hence the result is.

At least as far as their building activities are concerned, understanding termite behaviour is essentially about complex interactions. If you leave them out then you lose the essence of the phenomenon you are interested in. Termites need building material, they need the stimulus from other invdividuals and from their unstructured environment. Take that away and the termite does not make sense any more. With termites the Galileian trick of idealisation does not work. The termite itself is not sufficient to explain the termite mounds in the sense that it does not contain the essence of the system any more. Unlike friction and the wind in the case of free falling objects, other termites and the environment are not just disturbances to some "termite essence," but they are fundamental aspects of the system. Unlike the free fall, if you take away what complicates the system then you lose its essence. The interactions between termites and termites and their environments are irreducible.

Termites are also very different from the highly interacting molecules in a gas. In thermodynamics the particles interact highly, but the individual members of the gas, the molecules are undifferentiated. It does not matter whether Fred or Kate the molecules are here or there. One molecule is like the other. It is a billiard game with only white balls. Termites are not like this. They are spatially and functionally structured. Unlike the interactions of gasses that are homogenisers of the system, interactions in termites lead to more structure, both within the population and within the environment they inhabit. Symmetry breaking means that the insects start to prefer certain areas more than others, aggregating here while avoiding there. Ultimately, this is reflected in the shape and internal architecture of their mound. Not only that. Once the mound is there, then its structure in turn feeds back and modifies the behaviour of the termites. Within their mound they start to do things such as building a chamber for their queen or to farm fungus in special garden chambers that they build underground. To consider the average behaviour of the termite society is not informative at all. We would not be able to understand from it how the mounds are built. It is possible to describe an ideal gas in terms of its pressure, volume and temperature even though the gas may consist of millions of members. Describing termites in terms of averages misses the point.

When Bonabeau and Theraulaz wanted to put their ideas to the test they had to use novel methods. Traditionally biologists would engage in experiments. Get real termites into their lab and experiment with them under controlled conditions to find out whether or not their hypotheses are correct. This is not what Bonabeau and Theraulaz did, at least this is not all they did. They expanded their methodological armoury by building computer models of termites. They wrote a simulation model to encode the behaviour of the natural system, in this case the behaviour of the termites, how they can move about in their environment, where they go and what they do. In essence they create hundreds of *in silico* termites within their computers. Each of those simulated animals behaved according to the rules that Bonabeau and Theraulaz determined.

Unlike traditional equation-based approaches in physics, computer simulations can deal much better with interactions. In a sense, there are in-between traditional wet-lab experiments and a full mathematical theory. Certainly in the case of termite behaviour, computer modelling proved to be a powerful method, but it is not without its limitations. Simulations lack the generality of mathematical equations. They are only valid for a particular set of conditions. In Bonabeau and Theraulaz' computer model they had to specify every single detail of the system and then run the model to check what happens, then change the conditions a bit and check again and so on. This is the only way to find out how the system works as aspects of the model are changed. The kinds of things they may have varied are the intensity of the pheromone smell or the sensitivity of the termites to pheromones, the number of termites, the speed with which pheromone evaporates and so on. Needless to say that they could only sample a small subset of all possible variables.

They could not know what sorts of behaviours are produced by all the variant models they did not explore. Every simulation is only valid for the very particular assumptions that have been made for this very specific run. Beyond that nothing can be said with confidence.

Physicists tend to dislike this lack of generality. The shape of the termites' mounds depends on the environment and some very particular parameter settings. Change them and you may change everything. There is an inherent messiness in such simulation models. They are a far cry from the platonic beauty of equations in physics. Yet, the need of simulation models is just a reflection of the fact that there is no essence to termites in the same way as there is an essence to the free fall. When it comes to understanding the behaviour of termites, the details matter. This is apparently at odds with the established Galileian idea of physics. Consequently, physics has traditionally ignored complex systems. Not that termites are not interesting to study, but they do not allow general theories and hence are not loved by physicists.

Landslides, Sandpiles and Evolution

Termites would not have been so interesting if they had remained an isolated case. Yet, in the 1980s and 1990s scientists started to discover more and more systems that just like termites showed complex overall behaviour that can be reduced to simple rules. Shoals of fish are apparently governed by an invisible hand, in reality single fish follow simple rules: Keep a certain distance from the fish next to you; whenever the fish next to you changes direction, do the same. Again using computational models scientists could show that simple rules like this can account for the apparent magic of the shoal. The complex and organised behaviour of a fish swarm can be reduced to a few interactions. The same has been shown for ants and how they locate their food. A few simple rules about following a trail of pheromones and the ants can locate and exploit the best sources of nutrients in their environment. Joshua Epstein and Robert Axtell (Epstein and Axtell, 1998) tried the same trick for social science and economics. They devised simulations with simple agents that used elementary rules to exchange and trade goods. A similar conclusion emerged: Complex trade-networks are generated by simple rules.

All these examples suggested at the time that maybe the apparent complexity of the world can be reduced to a fundamental essence after all. Like Galilei's free fall in vacuum, but different nonetheless. Maybe all complexity can be reduced to some simple rules of interactions. Is this the Galileian approach 2.0? A new scientific revolution that finds the fundamentally simple behind the seemingly complex? A new method in science that finally extends the Galileian method of idealisation, taking it out of the laboratory into the world.

The similarity to the Galileian revolution is indeed striking. Just as the problem of falling bodies seemed rather intractable to Galilei's forebears because they did not do the right kind of abstractions and simplifications, so it seems nearly impossible to understand how complex structures such as termite mounds can be built. However, once the right abstraction is used, then the apparent complication boils down to simple rules of interaction. In this second Galileian revolution it is no longer the reduction to an idealised highly isolated system that reveals the simplicity of nature. In Galilei 2.0 this simplicity is revealed by a reduction to simple rules, that when applied within a realistic environment lead to complex results.

One thing is missing from all of this, however: Generality. Termite models are only valid for termites, fish models are only valid for fish and economics models are only valid for trade-networks. Even though the models are simple, they are extremely limited in scope. Compare this to Galilei's square law. It applies to falling termites, fish and traders, and anything else that falls.

This is the kind of generality physicists like. Naturally, the question was whether the same kind of generality can be found in these systems.

Progress came from an unexpected source. Sandpiles. The kind of thing children play with on beaches. Imagine the following: Drop a grain of sand to the ground, then another one at the same position, then another one and another and another and so on. Clearly this procedure, if repeated long enough will lead to a pile of sand first, then to a mountain of sand. To avoid growth without end, let us imagine that the grains of sand are dropped on a plate, a plate that is placed above an abyss of nothingness. Whenever a grain of sand falls from the plate, it simply disappears in the void. What we get now is easy to see. First, a pile of sand that grows, then at some point, the base of the pile has reached the edges of the plate and the pile will not grow any longer. As we drop new grains of sand onto it, other grains will fall from the plate and disappear into the abyss.

This is now where it gets interesting. Let us drop grains for a long time. We stop when the base of the pile has reached the edges of the plate. Now we change pace. Whenever we drop a grain of sand we pause and observe what is happening. Let us assume that there are no external disturbances, that is there is no wind blowing and nobody shakes the plate, or anything like that. We can now ask: How many grains of sand will fall off the plate when a new grain is added? One is tempted to say that a single grain will fall. After all, in the long run we expect the size of the whole pile to be constant, so as many grains must be lost as grains are gained. But this is only what happens in the long run. It does not mean that exactly a single grain falls for a single grain being added. Indeed it seems rather unlikely that for every grain of sand that is added exactly one grain of sand falls into the abyss. Sometimes a grain will just stay on top of the sandpile, or it will roll down the slope a bit before it gets stuck. The next grain to be added may still not fall down the void, but also remain on the heap. Indeed, it could be that many grains are added before eventually, one of them dislocates and tumbles down the pile. Maybe, in the process it dislocates other grains. It may even trigger a landslide where suddenly a large number of grains fall down the slope. Or it could just be the one that has just been dropped. We do not know.

The first person who considered this problem seriously was the late Danish physicist Per Bak. Instead of trying to write down equations to analyse the system, as one would have done in physics, he decided to make a computational model to find out what happens. Being a physicist he could not help but simplify the model. In real sandpiles grains come in all shapes and forms and they stack up in irregular ways. This is hard to model. Every single grain is probably different. Modelling different shapes is difficult. It does not only require that the shapes themselves need to be somehow represented, which is difficult enough. Bak would also have had to specify precisely how the various shapes interact, he would have had to calculate the friction between them and all the other forces that determine whether or not a grain of sand will drop or not. This was way too much trouble for Bak's purpose – he was only interested in the principle, not in real sandpiles. So he decided to radically simplify the problem. He assumed that sandpiles are cubes. Not only that, but he also assumed that there is only a discrete number of places where the grains can fall. The effect of this was that grains of sand could only be stacked up in nice towers, so that a grain never overlaps more than one other grain right beneath it. A rather unrealistic arrangement, but good enough for the purpose at hand.

In his model Bak used a square lattice (essentially a checker-board pattern) to arrange the stacks. Each square on this lattice can take a positive value, say 0, 1, 2 or any other value representing its height. It is in the nature of the geometry of these square lattices that each entry has 4 direct neighbours one to the east, west, north and south. (The ones at the edges of the lattice have some of their neighbours replaced by the abyss.) Bak now assumed that stacks become unstable when they are much higher than their neighbouring stacks. In this case then they fall. Real life sandpiles would fall chaotically into all kinds of directions. Again, Bak had to tame his electronic version of the sandpile and postulate a simple rule: Whenever a pile is higher than a given threshold value (say 10), then four grains are removed from the stack and one added to each of its four neighbours.

This rule is the central bit of the whole model. After one pile fell, it may now well be that one or more of its four neighbours has grown beyond the threshold size of 10, which means that it in turn has to fall, re-distributing its grains to all four neighbours. This in turn may push further sites over the threshold and so on.

In this way, each event of adding a grain of sand results in a cascade of topple events where stacks fall. In the trivial case the size of this cascade is 0, meaning that nothing falls. It could also be the case that these cascades are quite large and trigger many follow-on topple events. Bak's original question ("How many grains of sand fall from the pile when a new grain is added") is now transformed into a meaningful question within the model: When a new electronic grain of sand is added, how large is the ensuing cascade? He could of course not expect a specific number as an answer to this question because, clearly, each grain of sand will have a different effect. However, the great advantage of these computer models is that they allow the modeller to repeat experiments over and over again and collect a large set of data and obtain good statistics. Bak did precisely this. He counted how often he observed a cascade of size 0, of size 1 and size 2 and so on. He then plotted a graph where the x-axis is the size of the cascade and the y-axis is the number of occurrences. When he did this, he discovered that the distribution of the size of cascades follows a power-law. This does not sound very exciting, but it is. When you plot a power-law on a 2 dimensional graph and both graphs are in log-scale, then it appears as a straight line. This is probably the simplest form of dependence one can imagine.

Bak was excited about this, not only because of the simplicity of the straight line, but also because power-law behaviours indicated a deep connection of the sandpile to some well-known physics. These power-laws are often observed in phase transitions, such as the transition of water to ice or water to vapour. For Bak, who was a physicist, this was a revelation. It did not stop there. Once people started to look, these power laws popped up everywhere in seemingly complex systems. One famous example of a power law is the frequency of words in natural language. So, take a book (for example the one you are reading at the moment) and count how often the word "count" occurs. Then do the same for the word "how" and the same for "often" and so on, until you have counted every word occurring in the book. Then plot your data. As your horizontal axis use the number of words that occur once, twice, three times, four times and so on. On the vertical axis record how many words there are in the book that occur once, twice, three times and so on. For example, the word "Oryx" (which is a type of antelope in Southern Africa) occurs exactly once. For your plot you need to find all the other words that like it occur exactly once. Whenever you find one such word then you increase the count of words that occur exactly once by one. The word "antelope" occurs twice in this book. So, increase the count of words that occur exactly twice and move on to the next word. Say this next word occurs exactly 173 times. Now increment by one the count of all words that occur exactly 173 times and again move on to the next word. When you are finished counting all words plot the distribution of word frequencies, that is make a graph whose x-axis is the number of word-occurrences and the y-axis is how many words occur exactly x many times. The plot obtained in this way will, again, be a straight line on a log-log plot, at least near enough. Do the same for the frequency of earthquakes of a given size. You get a straight line in a log-log plot. Consider the size distribution of cities. You get a power-law as well. Look at the distribution of stock-market crashes of a given size. Again a power-law. Look at the size of extinction events during evolution, again there is a power-law.

All of these events, the writing of a book, the dynamics of traders on stock markets, the emergence of cities, the interactions of species in an evolutionary dynamics are very different processes with no apparent similarities between them and certainly no obvious relationships to known physical theories.

Bak's work seemed to suggest that maybe the occurrence of power-laws indicates a fundamental connection, a unifying principle between all these systems. If this was indeed the case, then it could be the beginning of the Galilei 2.0. A new revolution in science, the ground-breaking discovery that simplicity pervades not only the Platonian universe of theoretical physics, but the real world as well.

Perhaps, one may object, this initial enthusiasm is a bit overblown. Yes, the fact that there are power-laws appearing in all these different contexts is interesting, but it would surely go too far to conclude from this that there is a deep relationship between them. It could well be that there are fundamentally different processes that lead to the same distribution. Just because something is distributed according to a power-law does not mean that the power-law is of fundamental importance to the understanding of the system. It may well be that very different processes generate power-law behaviour. The onus is on the sandpile model to show that it is indeed relevant to all these examples. Bak took the challenge.

One of the systems he started to get interested in was evolution. The traditional way of looking into the past of biological history is to consider the fossil record and to see what kinds of creatures have existed during the various phases of the evolution of life. The idea is very simple. Creatures may get buried by all kinds of processes, either during their life or after that. They may, for example be so unfortunate to end their lives in a landslide, or they fall into a hole in the ground which then subsequently gets filled up, at least partially with some material. Marine animals may die, sink to the ground and be covered by sediment. In any of these cases it may happen that over long periods of time the covered organisms turn into rock, leaving a permanent record of its shape and form for us to study. The vast majority of organisms vanish from the face of the earth without leaving any trace of their transient existence. A successful fossilisation event takes a confluence of unlikely events and is exceedingly rare. Nonetheless, paleontologists found that these fossilisation events are frequent enough to give a fairly good picture of the diversity of life in the past. Using the rockified remains of former animals and plants and bacteria, scientists were able to understand at least partially how the organismic form changed during the eons of the history of life.

Darwinian theory, at least the way in which it is commonly interpreted, suggests that evolution is a smooth and slow process. In the fossil record this should manifest itself through a slow change of species. However, this is not what paleontologists see. Far from indicating a smooth change of forms and shapes, the fossil record appears to give evidence for long periods of stasis where nothing changed. These periods of calm are interrupted by intermittent periods of upheavals, sometimes large sometimes small. During these chaotic intermittent periods there are often extinction events where many species simply die out and new ones are created within the blink of a geological eye. Many of these extinction events are rather small, where only a few species disappeared. Others are catastrophic in size and a very large proportion of all species disappeared. There are enough of those extinction events to be able to analyse their statistics. And as it turns out, these extinction events are distributed according to a, wait for it, yes a power-law.

Naturally, paleontologists were interested in the causes of these extinction events. The small extinction events are easily dismissed as noise and are intuitively not so hard to understand. After all, this is evolution. New species appear, species die out. This is life. However, what about the large cataclysmic events? How can they be explained? One possibility is to assume that they were triggered by external events, such as large eruptions of volcanos or meteors that hit Earth, sudden changes of the climate or the like. For some of the events scientists were indeed able to find evidence for such external shocks. For others they did not.

Per Bak got interested in this, he thought that the extinction events that characterise evolution are a bit like the land-slides in his sandpile model. Probably extinction events are like the cascades of toppling grains of sand that he observed in his model. He started to suspect that there is a deep connection. Maybe the sandpile model can explain extinction events in evolution? This would certainly be an intriguing possibility.

Intuitively Bak saw the connection between his sandpile model and extinction events. To help the rest of the world see this connection as well he teamed up with the Danish physicist Kim Sneppen to make a variant version of the sandpile model that clarifies this connection. The model is very simple. Bak and Sneppen imagined species being arranged in a circle. Each species has exactly two neighbours to which they are connected. To have a proper model of evolution, each species also needed a fitness value (as in survival of the fittest). In the Bak-Sneppen model this is, initially, just a random number between 0 and 1. The next ingredient that is required is selection. To do this Bak and Sneppen located the species with the lowest fitness on the circle and eliminated it. Elimination was much less bloody in the model than it would be in reality. All it involved was to assign a new, randomly chosen fitness to the "victim." Bak and Sneppen did not leave it there. They reasoned that species never live on their own, but interact with one another. For example, lions need Zebras as prey. If Zebras die out, then lions will suddenly be in trouble as well. They incorporated this idea into their model in a very simple way: Whenever they mutated the lowest fitness species on their circle, then their immediate neighbours to the right and left of the latest extinction victim were assigned a new, random, fitness as well. Once this was done, they started from the beginning again and mutated the species with the lowest fitness on the ring. They repeated this over and over and over again.

At first the model seems trivially simple and it is hard to imagine that anything useful would come out of it. The simplicity is deceptive though. As it turned out, once the model had been running for some time, Bak and Sneppen started to observe cascades of extinctions. To understand this, remember that whenever an individual is mutated, then both of its neighbours are mutated too. This involves assigning a random number to those individuals. It may well be that either of the three mutated individuals again ends up being the lowest fitness individual on the ring. In this case it will be mutated again. A cascade is then a sequence of mutations that remain in one neighbourhood continuously. In this sense, an initial mutation of a species may create a whole cascade of subsequent mutations, or as Bak and Sneppen called it, extinction events. Bak and Sneppen collected statistics about the size of these cascades/extinction events and found that it again distributes according to a power law, just as the size of the landslides in the sandpile model or the size of extinction events in real evolution.

Following these initial models more and more sandpile-like models sprung up showing power-law behaviours in forest fires and gene regulatory networks and many more systems. All these models were inspired by the original sandpile model. A feeding frenzy started amongst the modellers. Eventually they had an inflationary effect. With so many models showing power-laws the question arose as to what they mean beyond the fact that they can reproduce some statistical peculiarities of the systems in question. What precisely can Bak's sandpile model, or the Bak-Sneppen model or any of the other models in the sandpile family teach us specifically about the real systems under investigation?

With Galilei's square law of gravitation, we know exactly what the model can do and when to apply it. With Einstein's theory of relativity, once we understand the formulas and how to apply them, it is also clear what they tell us. They make specific predictions about how time passes depending on the speed of objects. These theories can then be translated into experimentally verifiable predictions. Moreover, they also find practical applications. Based on the square law, Galilei could calculate the trajectory of bullets. Newton's laws are used as a matter of routine in physics to derive new laws. Einstein's theory of relativity is understood and applied regularly in technical applications, for example in GPS location systems.

The sandpile models and other complex system models never got that far. Many models have been published. Some better, some worse. Some more interesting than others. However, after decades of publishing models, there still is no synthesis in sight, no theory that ties individual models together. Something that exploits the apparent generalities, in a sense that is remotely similar to what physics has done. Maybe it is just early days. Perhaps we expect too much if we think that the Galilei 2.0 program will achieve in 20 years what has taken generations to complete after Galilei 1.0. Or, there is something fundamentally wrong with the idea of Galilei 2.0.

One problem is complexity. Galilei 2.0 is about complex systems and their essence. In the termite models, but also in the sandpile models the idea was to reduce the apparently complex system to a set of relatively simple interaction rules. These interaction rules are then the Galileian essence of the system, the idealisation that suppresses the noise and brings out what the system *really* is. Consider for example the Bak-Sneppen model.

The circle of species is a bare-bone representation of how real life evolves, at least that is the tacit assumption of Bak and Sneppen. It has to be the assumption, because otherwise they would not be justified in calling it a model of evolution. This assumption of an essence that is so crucial to Galilei 2.0 is hard to justify. When we think of evolution, then we think of a process that is inherently creative. The very hallmark of evolution is that it brings about creatures that find their own niche within a rich world. Organisms make their own living and they have the characteristics that allow them to deal with a very specific set of circumstances. Mango trees do not grow in England because it is too cold. Arctic foxes would struggle to make a living in the African planes. Evolution is about specific solutions to specific survival problems. The adaptive significance of organismic traits needs to be understood with respect to the very particular context within which it has evolved. The minimalistic Bak-Sneppen model represents none of this variety. It is essentially a game with numbers. There is no inkling of creativity there, nor a hint of the richness of the interactions in the biosphere. In fact there is nearly nothing in the model that is specific to evolution. If Bak and Sneppen had not called it "a model of evolution" it would have been difficult to guess for anyone that it was meant to be one. All it does is to reproduce some statistical distribution of extinction events. This is as far as the similarity goes. Extinction events are certainly important for evolution, but its statistics is hardly its salient feature.

It seems that the reasoning went upside down in the Bak Sneppen model. In the tradition of the Galilei 1.0 program, the idea is to identify the essence of a system (free fall in a frictionless environment). Based on an ideal system, one can then make predictions. If these predictions can be verified experimentally, then one can feel confident that the idealised system, the abstraction, is indeed a relevant one. In the Bak-Sneppen model and for that matter in many of the complex system models, the reasoning is inverted. They start with a model that reproduces some statistical features of the system. The statistical similarity then justifies the conclusion that that the model captures the essence of the system. Seen in this way, it is clear that the whole approach is flawed. Statistical properties can be produced by many mechanisms. The fact that two systems share the same statistical behaviour does not mean at all that they share some fundamental similarities. Hence, the fact that the sandpile model reproduces the statistics of extinction events or earthquakes means nothing at all.

The model and the real system are separated by a huge complexity gap. This gap cannot be bridged by the Galileian idealisation trick. It remains a wide scar on our body of knowledge.

Complexity and Biology

The legacy of Monod was not only a particular method to do biosciences, but also a research culture. Since the days of Monod biologists have developed a certain obsession with details, not quite pathological but not too far away from that. No biologist is content with finding out a bit about her organism, but they always want to know exactly how it ticks. Figuring out these details is not cheap or easy. Entire research careers are spent on decoding tiny aspects of a particular regulatory circuit of *E.coli,* or to know exactly what protein X is doing. Nearly every shred of biological knowledge we have represents many years of work by a team of talented biologists. Given the enormous effort it is no wonder that biologists developed a strong appreciation for details. This is not a bad thing in and by itself. However, there are side effects. The daily grind and pressure of a bioscientist's life leaves little time left to ponder on the big questions.

Is there an alternative biology, another way to understand organisms that complements the flat-pack factories into which bioscience research labs have now turned? It would perhaps be too much to expect that an alternative to flat-packing emerges from within biology itself. Biologists have too heavily invested into details. It is therefore perhaps not surprising that the first attempts at an alternative came from a different corner of academia. The pioneers who tried to come up with a non-flat pack understanding of biology were mathematicians and physicists. They had a crucial advantage over real biologists. They knew nothing or at least very little about the complex details that biologists have so painstakingly learned about. Blessed with ignorance, and with an acquired distaste for the particular, these outsiders took a very different approach to biology. Instead of asking what a *particular* organism such as *E.coli* is made of, they asked a different question: What is it about organisms *in general* that allows them to function. This is a question that biologists only ask, if at all, during cocktail parties at scientific conferences. The scope of the question is too broad to make a meaningful research program within the Monodian paradigm. Yet, mathematicians and computer scientists are a different breed of people. They are not interested in the biochemical particular of the *lac* operon; maybe they have not even heard about the *lac* operon. They want to know about fundamental, unchanging principles of how organisms work.

For many of those mathematicians and computer scientists it is natural to think of organisms as machines. At a first glance, an organism may not seem so very different from a machine, such as the Visilab car. A closer look, however, reveals that they have some properties that machines do not usually have. The most obvious one is that they multiply. Take one cell of *E.coli,* add a bit of nutrient and you will end up with two. Try the same thing with the Visilab car, or any other man-made machine; the trick will not work.

The second remarkable property of organisms is that they are able to repair themselves, if they are damaged. At least to a certain extent. Broken bones grow back together again, cuts in the skin mend and bruises heal. Try the same thing with the Visilab car and you will be disappointed. A broken suspension must be changed. Putting it in plaster for a few weeks is not sufficient.

The third property is evolvability. Not only can organisms propagate, but they can also acquire hereditary modifications. It seems that organisms have an internal plasticity that, at least over generations, can lead to significant modifications of their *bauplan*. The ability for change is quite remarkable. Over admittedly long time scales, but still, small changes from one generation to the next have led from the first simple organisms to both the reader, *E.coli* bacteria, the grass on the lawn and many thousands of other types of living systems. More impressive than the plasticity itself is the apparent openness of the process. Rather than changing a bit before eventually settling onto a final optimally adapted form of life, it seems that natural evolution has remained creative since its beginning, constantly coming up with new life forms, creating new niches, undergoing major and minor revolutions in the process.

While the Galilei-Monod biology is getting lost in details, mathematicians and computer scientists started to ask about the properties a system needed in order to display these properties. Few of them had a formal training in biology. This enabled them to leave the well-trodden path and explore biological systems from new perspectives.

Self-replicating Machines

One of the first researchers who got interested in computer models of biological systems was the Austrian-Hungarian pioneer of computing science Johann von Neumann. In the 1940s, some time before Monod, he started to think about how one could construct machines that reproduce themselves. However, being a computer scientist he was not so interested in physical machines, but computer simulated ones. Software entities that were, somehow, able to make copies of themselves.

In a sense, copying things is trivial in computers. On modern machines a simple combination of keyboard shortcuts and clicks and drags and drops is enough to make as many copies of a file, or indeed any software object as one likes. With a tiny bit more effort this process of copying can also be automated. For any first year computer science student it is easy to write a program that can make a copy of itself and reproduce in this sense. However, somehow simple copy operations of this sort seem overly trivial when compared to reproduction in biological systems. This feeling of dissatisfaction is reinforced when we consider in more detail what happens when a copy operation is performed in a computer. Upon issuing a copy command, the operating system will first determine the size of the file to be copied. Dependent on this, it will then allocate an area on the hard-disk (or whatever storage medium is in use) where the copy of the file is to be stored. Once this is done, it will then read the file to be copied bit by bit and write the information into the newly allocated space.

The details of this process naturally vary with the operating system and the storage technology that is used. The important aspect, however, is not the technical detail, but the fact that copying a file requires the external machinery of the computer. When files are copied, they are copied by the computing hardware that does the leg-work.

This is very different from the way organisms, say *E.coli*, replicate. When the bacterium makes a copy of itself, it does not merely issue commands to some external central processing unit, or some external Grand Copier to make another *E.coli*. Instead, the bacterium has internal processes that through their collective action lead to the creation of a new cell. This is not to say that they are independent of their environment. Bacteria and other organisms crucially rely on an adequate supply of nutrients and the right environments. Ultimately they also need the basic physical laws on which the agency of their components is based. Yet, there is no Grand Copier that replicates the cell atom for atom.

When von Neumann set himself the task of creating a self-replicating machine in software, he was not interested in merely copying programs. His idea was to create an artificial world *in silico* in which objects can replicate themselves in a non-trivial way. The idea is that these "objects" consist of a number of simulated components that are the elementary particles of the virtual worlds. Any properties of these objects are equally a consequence of their particular composition. In this setting one can think of a non-trivial replicator as an object that can make copies of itself, while none of its constituent parts by itself has the ability to replicate. Real *E.coli* cells are non-trivial replicators in this sense; their constituent parts cannot replicate on their own, but the cell as a whole can. Von Neumann wanted to know what sort of properties an artificial world had to have in order for it to support self-replication.

In his search for a simple way to create such artificial worlds, von Neumann came up with the idea of a *cellular automaton*, or simply CA. The basic unit of a CA is an automaton. Such an automaton is an abstract entity that can take, at any one time, one of several states. It also has one or several input channels over which it can receive "commands." Whenever it receives a command it goes into a new state. Precisely what this new state is normally depends on the commands it received but also on its present state. The new state could be the same as the old state.

The simplest automaton has two possible states, say "on" and "off." For simplicity, let us assume that it has exactly one input, and the input can only be one of two commands. This simple case is then a bit like Uncle Herbert's moods on Sunday afternoons. Uncle Herbert is a big fan of basketball. For him the highlight of the week is to watch the weekly basketball game on TV, following his favourite team. Unfortunately, his Sunday afternoon mood can depend very much on the outcome of the game. If he is in a bad mood on Sunday and the team wins, then his mood improves and he will be in a good mood for the rest of the day. If he is in a good mood anyway and the team wins he remains in a good mood, but if he is in a good mood and the team loses, then he gets into a bad mood. Likewise, if he is in a bad mood and the team loses he remains in a bad mood for the rest of the day.

Uncle Herbert's moods can be described very much like an automaton. His own state (his mood) changes in response to input (the outcome of the game), but also on his previous state (the mood he was in before watching the game). The simple "Uncle Herbert" automaton can be described in a table as follows:

Game result	**Previous mood**	**New mood**
win	good	good
lose	good	bad
win	bad	good
lose	bad	bad

Instead of "win" and "lose" we can now use "1" and "0;" likewise, we can write "1" for good and "0" to describe a bad mood. Then we get a new table which is exactly the same as the old table, only that we have replaced words by the numbers 0 and 1.

Game result	**Previous mood**	**New mood**
1	1	1
0	1	0
1	0	1
0	0	0

Finally, to get to a proper description of an automaton, we replace "Game result" by "Input" and "Previous mood" by "State" and "New mood" by "New state."

Input	State	New state
1	1	1
0	1	0
1	0	1
0	0	0

This is a simple automaton or rather a state transition table describing the update rules for an automaton. There are many ways in which it could be extended. One could, for example, add more states. Uncle Herbert could be in a very bad mood or he could be upset or any other differentiation of mood. One could also have additional inputs to modify the mood table. Indeed, one can do whatever one wants, as long as one specifies in the rule transition table precisely how input commands lead to new states.

Just for the fun of it, let us now make an automaton with 2 inputs. All we need to do is to specify how a specific combination of inputs, together with the previous state leads to a new state. Table 3.1 shows one possible automaton. Its rules are very simple. It will always flick its state when its two inputs are the same. Many other rules are possible.

Input 1	Input 2	State	New State
1	1	1	0
1	0	1	1
0	1	1	1
0	0	1	0
1	1	0	1
1	0	0	0
0	1	0	0
0	0	0	1

Table 4.1: An arbitrary 2 state 2 input automaton.

The idea of automaton was already very well known to von Neumann. One of his (many) contributions to computer science was to take the next step that turned the abstract idea of an automaton into a powerful, if very simple imagined world with its own laws. He extended the idea of an automaton by considering interacting automata. To do this he arranged the basic automata in lines or square grids. This sort of spatial arrangement allowed for the possibility of connecting automata with one another, in the sense that the state of one could be the input command of others. For example, if the automata are arranged along a line, then each of them would have a neighbour immediately to the left and right. In this arrangement one could then stipulate that each automaton has three input commands, namely the states of the respective neighbours and its own state. Similarly, if they are arranged on a square lattice then one could come up with rules that relate new states to the existing states of the neighbours to the south, east, west and north, or the new states could also depend on the diagonal neighbours. One can choose freely. These collections of automata are precisely von Neumann's CAs.

There are no limits as to how one can organise a CA. One could, for example, have different rules for each automaton in the arrangement. This is absolutely possible, but in practice difficult to handle. To make his life easier, von Neumann therefore assumed that the individual automata in a CA have the same rule tables. Before he could do anything with the CA, he needed to specify first when and how the update rules of the individual automata are applied. Again, there are many possible ways to do this, and it seems wisest to choose the simplest method. He therefore decided to update the rules of all automata at the same time. In practice this means that for each cell the next state is determined before the cells are updated. So, the next state of the first cell is determined, but the state is not yet updated. Then the next state of the second cell is determined, but not yet changed.

This is done for all automata in the CA. Only when the next state of the last cell has been calculated will the new states be realised. This seemingly cumbersome procedure ensures that the order of update does not matter, because each cell only "sees" the old states of its neighbourhood.

Figure 4.1: A screenshot from a re-implementation of von Neumann's automaton with the software Golly. This shows only a small part of the entire configuration of von Neumann's replicator. Each shade of grey corresponds to a state.

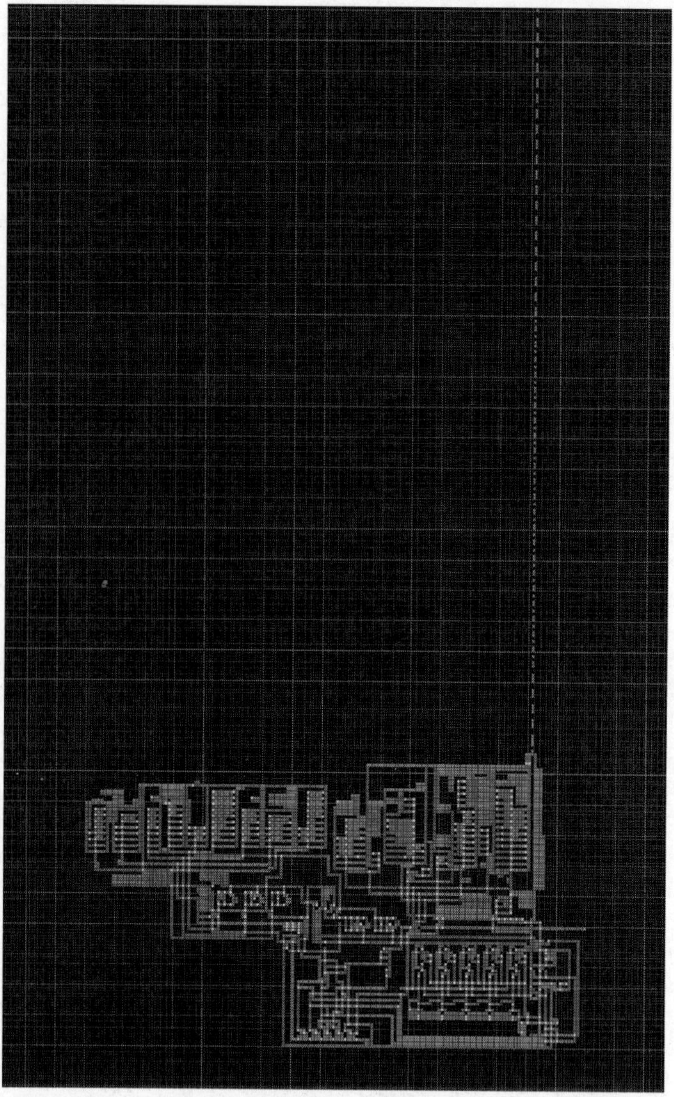

Figure 4.2: A recent implementation of von Neumann's self-replicating machine by Pesavento (Pesavento, 1995). The dark background is all dormant cells.

The crucial question that von Neumann faced next was in what sense a CA could reproduce. One possibility is to assume that CAs reproduce like bacteria. They somehow grow and after some time one CA becomes 2 CAs become even more CAs. In the context of computer programs—and CAs are essentially computer programs—this concept of reproduction makes little sense.

Instead of thinking of the CA as a whole as the reproducing entity, he was interested in objects within the CA that would reproduce. The CA itself is then a "universe" that supports self-reproduction. Consequently, he needed a conceptual notion of an *object*. In order to be able to distinguish "thing" or pattern from "empty space" in the CA, it was necessary to somehow define what "empty space" is in the CA. This is easy. Just define a dormant state called "0," and give it a special rule. Whenever dormant cells are surrounded by other dormant cells they would remain dormant, just as empty space surrounded by empty space is not doing much. Within this background of cells in state 0, von Neumann could then (manually) create an initial pattern of states other than the dormant state. This provides a natural definition of "things" in the universe as patterns that are separated from one another by empty space. If an initial pattern after many applications of the update rules results in a copy of the initial pattern as well as the original pattern, then this patter could be said to be self-replicating within the particular world of the CA.

The particular pattern von Neumann had in mind was a very peculiar one. His replicating "machine" as he called it was not a random configuration that he simply stumbled upon, but a highly structured arrangement of patterns. His replicator consisted of two major parts. Each of those two parts was essential for the functionality of the self-replication, but could only realise self-replication when they came together.

The first component was a *universal constructor*. This universal constructor is a pattern within the CA that takes descriptions of patterns as input. Once it receives such an input it then realises this pattern within the empty space surrounding itself. To understand the conceptual importance of this universal constructor, it is important to remember that it is itself a pattern of cells. Its ability to accept inputs and interpret them as descriptions of patterns is a result of the concerted action of hundreds of cells all with identical rules, but initialised to different states. None of these cells can by itself interpret patterns or construct anything. Only when the cells are arranged in just the right way in a CA with just the right rules, only then can the pattern function like a universal constructor.

The second component of von Neumann's self-reproducing "machine" was the description of patterns itself. This is comparatively easy. Von Neumann conceptualised this description as a tape, not a physical tape, but a string of states that are interpreted by the universal constructor as a specification of another pattern. With the tape and the universal constructor together von Neumann had a machine that could "read" a description of a pattern and produce this pattern in its vicinity.

The work was now almost done. All that was left to get a self-replicator in his CA, was for von Neumann to use a standard trick in computer science: He supplied the tape with a description of its own pattern, that is a universal constructor including a description of a universal constructor. Now the universal constructor would, following the instructions on the tape, produce a new universal constructor somewhere in the empty space of the CA. It reproduced.

Unfortunately, the CA von Neumann proposed was very big, consisting of several hundreds of thousands of cells and his CA had 29 different states. For many years, decades, von Neumann's CA stood as the first successful attempt at producing self-reproducing systems. In many ways he was ahead of his time. For one thing, the design he came up with was much too complicated to be implemented given the technology available to his contemporaries. Von Neumann himself died before finishing the design of his CA and his self-reproducing machine was later modified and completed by others, although it was never simulated in its entirety on real computers until fairly recently.

Even though von Neumann's contribution was not directly fruitful in the sense that it had led to any specific insights about self-reproducing machines, it inspired later researchers to find radically simpler self-replicating patterns in CAs. A reason for the size and complexity of von Neumann's CA is that he insisted on his replicator being a universal constructor meaning that the machine could construct any pattern that is given to it. This is a rather strong requirement that is certainly not met by real organisms. Living systems normally have a very limited ability to generate patterns. They can reproduce themselves and similar shapes. In this sense we can think of a mother's womb as a constructor for human babies. It can do some variations of the baby-theme, but would fail to construct, say, a crocodile, let alone a car. Requiring a universal constructor seems therefore to be requiring too much as long as one is interested in biological systems.

Figure 4.3: A detailed view of the Langton Loop

There was little activity in the field of self-reproducing CAs after von Neumann, until in 1984 the American computer scientist Christopher Langton came up with a radically re-thought self-replicating structure. He abandoned the requirement for a universal constructor and managed to construct a CA that is easy to implement and replicates within a reasonable time. Langton used a CA with 8 states and the self-replicating pattern is smaller than 100 cells, which is very small compared to von Neumann's universal constructor.

The initial pattern is a small loop, that looks a bit like a snail. Essential to its operation are the guiding sheath cells that protect the inside states from the environment. Within the sheath cells are the active cells that drive the self-reproduction of the initial pattern. When the CA is run with the correct initial pattern, then the head of the snail seems to extend forward at first, then it bends and extends leftwards forming a snail-house. It then closes in on itself, severs the connection to what used to be its house and forms a new snail-house. The circle can begin again.

Langton's loops can only continue to reproduce as long as there is free space in the environment, but it will stop once the space is filled. As a consequence each loop can produce at most four offspring; then there is no more space around it. From then on the loop is dead and remains static until the computer is turned off. Figure 3.7 shows two screenshots of these loops. The picture on the right hand side shows the Langton loops after a number of replication events. It is quite apparent from the picture that not much is happening any longer in the centre of the CA, although new snail houses are created at the edges.

The motivation for Langton, von Neumann and the countless researchers who followed up on their work was not primarily, or indeed at all to understand real life and how it reproduces, but to understand the underlying logic of self-reproducing artificial life forms inhabiting newly constructed worlds. They wanted to fabricate digital creatures, or life-as-it-could be. In many ways,

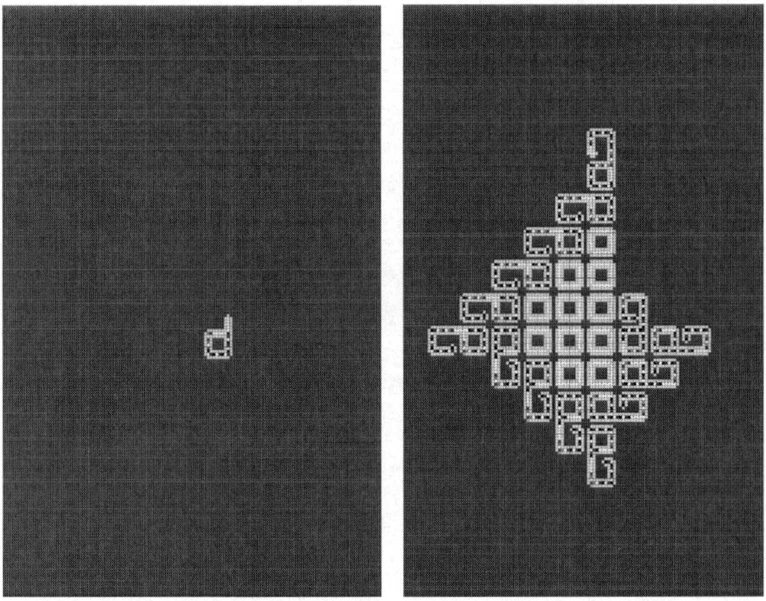

Figure 4.4: The Langton Loop in action. The initial pattern of the loop (left) and then the filled space after many updates where much of the space is now filled (right).

their results were impressive, certainly they have captured the imagination of countless computer scientists and mathematicians. The patterns seemed to be bare-bone models of the logic of self-replication. One could think that they are the life equivalent of Galilei's square law. Take away all the detail of real systems and boil is down to its bare minimum, then, so the tacit assumption, you end up with something like Langton's loop.

Sounds great. However, once one gets over the first excitement it becomes obvious that these self-replicating CAs are little more than patterns that multiply on a screen—indeed they *are* nothing but patterns on a screen. It is not at all clear what and indeed whether anything fundamental can be learned from them about biological systems. It is not even clear that the multiplying patterns are even separate entities. The patterns that seem to replicate on the screen are entirely a creation of our mind. We fool ourselves into seeing replicating objects on the screen, when indeed all there is to see are state transitions. There is no reason to see the pattern of grey-scale points as an entity, separate from the large background of dark-grey dormant cells. The entities are simply a figment of our imagination induced by a pretty visualisation. Remember that the internal "physics" of CAs does not allow for individual cells to move across the screen. What appears to move are but propagating state changes. There is no sense in which objects can move in CAs.

Seen from this perspective, these "replicating entities" in von Neumann's and even more so in Langton's CA are not entities at all, but rather cognitive illusions. Yet, if they are not entities, then they cannot have any agency. If they have no agency, then we certainly cannot learn anything about the origin of agency in living system about them. "Self-reproducing" CAs are little more than computational amusements.

This criticism applies perhaps to a lesser extent to von Neumann's replicator, which after all is a functional structure. The universal constructor can be interpreted as a pattern that has some intrinsic computational properties. When fed with a tape describing a pattern, then this structure can produce this pattern. In this case self-reproduction is a mere trivial corollary of the universal construction. Once you have a universal constructor, then it is impossible not to have a self-reproducer as well. Von Neumann's CA is therefore an impressive instantiation of a non-standard computer, a spatially distributed pattern plus rules that can compute other patterns. Real living systems are not universal constructors and do not need to be universal constructors, nor are they interpreting tapes with descriptions in order to self-reproduce. Therefore, von Neumann's CA has little to teach us about life.

Self-repair

Some may be convinced by the self-reproducing CAs. On the

other hand, if you are not naturally seduced by the pattern replication in CAs, then it will be hard to convince you that this is a proper model of self-replication. I doubt that any biologist will take those computational amusements seriously. It is unlikely that anybody really can learn something about self-reproduction and its logic by studying von Neumann's universal constructors or Langton's loops. Their whole approach leaves a strong sense of dissatisfaction.

One of the main ingredients missing from these CAs is a natural way to delineate an object from its environment. The idea of dormant cells tries to address this, but it does feel artificial. There is no natural barrier that allows entities to remain separate in CAs. Soon, scientists started to think in more depth about what it means for something to be a separate entity and tried to come up with clear concepts.

One important idea that some felt could be useful was the concept of *autopoiesis*, originally due to the Chilean neuroscientists Umberto Maturana and Francisco Varela. Maturana and Varela noticed that life is all about exchanging matter with the environment. This exchange enables growth and replacement of damaged components. Once the processes that keep the constant turn-over of matter going are disrupted, the organism itself dies. It ceases to exist and becomes a lump of matter. From this very simple observation they concluded that the material composition itself is not important for living systems. What counts is the functional organisation, that is the ability of organisms to keep this constant turnover of components going, to find nutrients, to process them and to self-repair. If this continuous turnover of components is so crucial for life to continue, Maturana and Varela reasoned, then it is clear than it must be true for any life form that each of its components is (i) either available as a nutrient from the environment or (ii) it can be synthesised by the organisms using any such nutrients.

When the ability to continuously produce these components

is crucial, there is another even more fundamental property of organisms that is necessary. Organisms need to be well separated from the rest of the environment. They need to be distinct. It must be clear which part of the world is the organism and which part of the world is its environment. If this was not the case, if there was some sort of ambiguity about where the living system begins and where it ends, then the very concept of organism would not make sense any more. A system whose boundaries are unclear could not synthesise all of its components, it could not reproduce or repair itself, because it would not be clear what precisely it is that needs to be reproduced and repaired. Maturana and Varela concluded that organisms need to have some sort boundary that clearly differentiates them from their environment. They did not think of this boundary as a seal, as an impenetrable boundary, but more as a semi-permeable membrane that demarcates a living system from its non-living surroundings and still allows exchange of material.

Maturana and Varela called systems that had both a boundary and the ability to turnover components *autopoietic* systems. The word autopoiesis is Greek and means self-building. They postulated that autopoiesis is a fundamental defining property of living system. The Visilab car is clearly not an autopoietic system. It does have some autonomy and it is perhaps clearly demarcated from the non-Visilab rest of the world. However, the car cannot produce its own components. If the exhaust pipe breaks, then a new one needs to be manufactured in a factory and then fitted by an engineer. The same is true for the gear-box and the engine pistons and virtually every other component of the car. Contrast this now with an *E.coli* cell. Give it some nutrients and it will produce the rest of its components, to the extent that it can replicate. Self-reproduction means that the cell has somehow been able to synthesise all of its components given only the nutrients in the environment.

The Langton loops above are not autopoietic systems. They do not have internal factories that produce parts, which explains to some extent why they appear to be trivial and ultimately unsatisfactory. The case of the von Neumann CA is a bit more complicated, because it seems to have the ability to produce parts of itself in a non-trivial way. However, in the world of von Neumann's CA there are no objects, no components that can move, only states that can change. Consequently, there cannot be any turnover of components, because there are no components. The CA seems entirely static, more like the Visilab car than like an *E.coli* cell. Moreover, von Neumann's self-replicator fails to be autopoietic because it lacks a suitable boundary. It is not protected from its environment and it is not sustained by selective uptake of external matter. This makes it very vulnerable to interferences with external objects.

Maturana and Varela thought that it would be useful to demonstrate the concept of a autopoiesis in concrete terms. A tangible model of a simple autopoietic system that could focus the discussion and clarify the theoretical concept. To do this they came up with the so-called SCL model. This model is a very simple imagined universe, an artificial chemistry if you will. It consists of a discrete 2D space in which chemical species can move and react with one another following rules that Maturana and Varela specified. Concretely, there are three types of molecules, namely **S**ubtrate, **C**atalysts and **L**inks. The Substrates are molecules that are meant to represent the food or nutrient that the cell receives. When two of these Substrate particles come into close spatial proximity, that is if they get into each other's neighbourhood, then they can be converted into a Link particle. However, this only works if there is also a Catalyst particle in the neighbourhood of the two Substrates. The opposite reaction can also occur. From time to time a Link will decay into two Substrates. The Link particles themselves play an important role in the setup. As their name suggests they can form bonds with one another. Each Link can bond to at most two other Links.

Again, the bonds between Links were allowed to break from time to time, in an entirely stochastic way.

The ability of Links to form chains is a core part of the model. A closed chain of bonded Links forms the shell, so to speak, of the simulated autopoietic system. It represents a membrane, the bit of the minimal simulated cell that distinguishes its inside from the rest of the world. The rules were also set up in such a way that the closed chain of Links prevented Links and Catalysts from diffusing into/out of the cell, but the Substrate particles could freely pass through Links and hence through closed chains of Links.

Substrates move more or less randomly through their grid-world. Whenever a Substrate particle diffuses through a cell barrier then it can be used as repair material by the proto-cell. Maturana and Varela also included a Catalyst particle within the cell. Note that the Catalysts were not able to penetrate the membrane as long as the closed membrane chain was intact. Hence the Catalyst can be trapped within the enclosure of Links. Now remember that two Substrate particles are converted into a Link when they collide with a Catalyst. Within the relatively small area that made up the autopoietic cell, it was likely that this would happen.

This in itself is not the interesting feature of the model. If Maturana had placed a Catalyst anywhere outside the cell, it would also have started to produce Link particles. In this sense there is no difference between producing them within or without the cell. Except that the membrane constrains the movement of the Catalyst and of any Link particles. The concentration of Links within the cell will therefore be very high, simply because Links are continuously created inside but cannot escape from the membrane. This is critical for the entire model.

Bonds between Links forming the membrane will break from time to time. Sometimes a Link within the chain will disintegrate into two Substrates. Whenever this happens the membrane breaks. However, this break may be fixed. Often there will be free Link particles available near the site of the break, ready to form a new bond with the free ends of the membrane. If that happens, then the membrane is repaired. This can only work if the concentration of Links inside the "cell" is high. Maturana and Varela could show that this is an effective mechanism, in the sense that it extends the lifetime of the closed chain of Links well beyond the lifetime of its individual components, the Links. However, it does not work indefinitely. Sometimes, a Link breaks and then the two ends of the membrane move far apart before the bonds can be mended again. In this case, self-repair fails, the

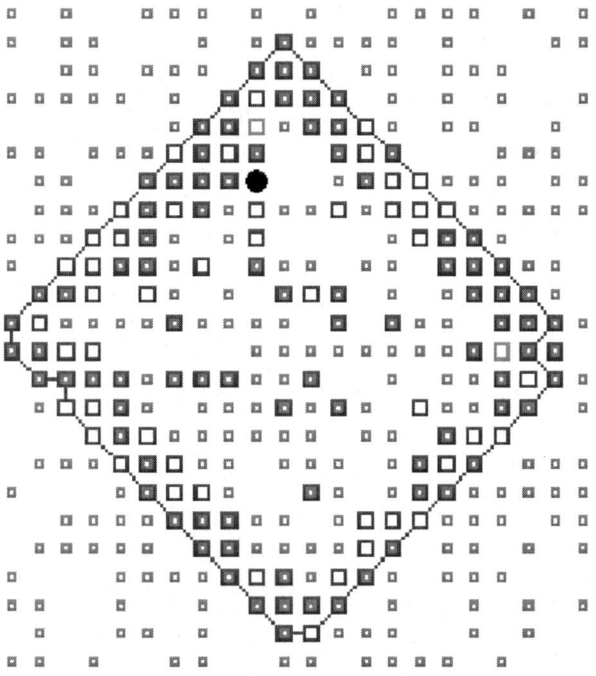

Figure 4.5: A membrane in the SCL model. The Catalyst is the black round dot, the Links are the large squares and the Substrate particles are the small squares. The closed chain of Links is clearly visible. Note the Links close to the inner side of the membrane. If the chain ruptures there will be a Link ready to step in and repair the membrane.

Catalyst can escape and the structure disintegrates quickly. Death in a computer model.

The SCL model is nice. No doubt. It has a distinctive biological feel to it, at least in comparison to other conceptual computer models. If one has already bought into the concept of autopoiesis, then the SCL model may be quite interesting. If, on the other hand, one is not a follower of autopoiesis or at least if one tries to apply the SCL model to understand a specific real biological systems, then it is far less clear how to interpret the SCL model. What is it that we can learn from it? One possibility is to take it as a model, or rather an implementation of a minimal cell *in silico*. If we follow this, then we would have to interpret the closed membrane as an example of a minimal cell implemented in a computer. If autopoiesis is indeed a sufficient condition for a system to be alive, then we could think of the SCL membrane as more than a mere model of life. It would be a computational instantiation of life. The problem with this interpretation is that life would become cheap indeed. The agency of the SCL membrane is very impoverished. The membrane is not doing much. The "mechanism" that keeps it alive is very simple and hardly deserves to be called "mechanism."

It is clear that Maturana and Varela, when they came up with their SCL model, were inspired by the semi-permeable membranes of real systems and how they maintain a very specific chemistry within the cell. However, in their model there are no special processes going on inside the proto-cell. Anything that can happen within the membrane can also happen without. The chemistry of the proto-cell is not different from the chemistry of the surrounding area in the same way as the chemistry within an *E.coli* cell would be very different from that of its ambience. The autopoietic system of the SCL model does not have any sophisticated bio-synthetic pathways that convert nutrients into biomass. The function of the membrane is limited to that of a barrier that increases the concentration of Links inside. Nothing more. A poor cartoon.

If we believe that the SCL model is indeed an implementation of a living system, a digital creature, then this means that the notion of "life" that underlies it is very wide in scope. Ultimately we can define anything in any way we want. We can call our bread caviar. Then we all have caviar for breakfast every day, but this does not make us rich people. By the same token, if we call the chain of Links a membrane, that does not mean that this chain of Links does indeed have essential features of living systems. If we take a step back from the model, it appears ridiculous to think that the SCL model is alive. It is not. It is just a very simple computer simulation. Nothing more.

At the same time it has some undeniable appeal. There is something intangibly interesting about it. I am inclined to think of it as a basis of discussion. What it does is to challenge us to think about what precisely we mean when we say that a system is living. Does it need to be made of carbon or can it be made of bits and bytes alone?

When we ask this question, what we really ask is whether there is any essence to living systems. Is there a minimal set of conditions that needs to be met for something to be alive? Can we formulate some abstract model of living systems that on the one hand avoids the trap of the Monodian flat-packs, but at the same time provides an elegant formulation of the essence of life? It seems to me that this question is what really underlies the attempts by Maturana and Varela and all the thousands of scientists that came after them with their own models. They all engaged in a quest for the Galileian essence of life, something fundamental that transcends the particulars of individual living systems. Something that helps us to escape from the information black hole in which we were forced following the Monodian science.

This quest for the essence of biology is still ongoing and it has been largely ignored by biologists. They are not interested in it at all. The main reason is precisely the lack of specificity. Even if one day a scientist came up with a computer model that demonstrated the fundamental nature of life, it would not tell the biologists anything about the mechanistic details of how the lactose system or any other specific system works. The model would not be biologically specific and therefore it would probably not receive much love from professional biologists. Here we find another contradiction. As long as one is specific, it will be flat-pack science. The moment one tries to uncover fundamental and general principles in biology then the models lose touch with reality and remain irrelevant. The Galileian trick does not seem to work in biology.

Evolvability

Both the SCL model and the various attempts to design self-reproducing CAs left out one aspect of real life that is often thought to be essential for life: Evolution. All the variety of life we see today has evolved from our primitive ancestors. The rhinoceros, the fruit-fly, sharks, whales, the apple tree and everything else ultimately traces back to some common ancestor many billions of years in the past. What the first proto-life looked like precisely, nobody knows at present. It is generally thought that it was some sort of chemical system that was somehow encapsulated in a membrane and was able to make more of itself and its membrane.

This initial ur-form of life was probably not very efficient at making more of itself. At some point, somehow, the chemical composition of one of these first proto-life forms would have changed. Maybe it just changed because of some random variations that happened during the process of reproduction, or somehow a new type of molecule diffused through the membrane into the cell. Or maybe there was some other sort of variation. Whatever it was, as a consequence of one of those changes the ur-life form reproduced faster than the first protoreproducer. Maybe, it did not even do that, but it was more reliable or more stable. In any case, it had some advantage over the old form that allowed it to outcompete its ancestor.

And so it went on. This second type of life was supplanted by a third which was then further improved on by a fourth and so on. Change continued from then on and entire eco-systems composed of many varieties of life emerged. Ultimately this process resulted in you and me as transient links in the chain of life.

Some think that this process of continued change in the face of competitive interactions in the biosphere is what characterises life. The mechanisms that bioscientists study, the wolves and mountain foxes, even the scientists would not exist if there was no evolution. One of those who felt that evolution is the key to understanding life is the American biologist Tom Ray. Looking at all the previous attempts to model life in computers, he felt that evolution was missing in these models. He also felt that one should perhaps take the study of life using computers to the next level. Rather than trying to implement organisational forms that have evolved for real carbon-based life, maybe one can get new insights by studying computational life-forms. His idea was to drop any pretence of modelling what is going on out there in the real world and to create novel forms of life that are perfectly adept to the binary universe of a digital computer.

Ray reasoned that a genuine computational life form would be a computer program. Not just any program, however, but a program that can reproduce itself, so a program that can make copies of itself. Note here the shift in thinking away from real biology. Langton and von Neumann were both very concerned with the problem of trivial re-production in computers and spend quite some effort to construct computational universes that could support entities that somehow had the ability to reproduce themselves while not simply being copied. Ray, on the other hand, did not worry about it. His new idea of reproduction is a computer program that could copy itself, using the external machinery of the computer. For Langton and von Neumann this would have been a case of trivial reproduction. Not so for Ray.

Writing a computer program that can make a copy of itself is not particularly challenging. Every programmer with a little bit of skill can achieve this. However, Ray did more than that. He wrote computer programs that could evolve. In order for that to be possible, self-reproduction is essential. However, another crucial ingredients is needed: Faults, or mutations as they are normally called. Real organisms make mistakes when they reproduce which manifests itself in a slight change of the genome. Often these changes are inconsequential, sometimes they are deleterious and occasionally they make the organism better suited for its environment. Mutations drive evolution.

Computer programs that can copy themselves, normally do so perfectly or not at all. Yet, Ray wanted to implement evolutionary effects into his self-reproducing programs. To achieve this he had to artificially specify that from time to time a program makes mistakes when copying itself. The offspring of this self-reproducing program would then be somewhat different from the original. His idea was that these mutations would occasionally produce a program that copies better than the parents and outcompetes them. Concretely, he hoped that if he started with a hand-designed ancestor self-reproducing program, over time new programs would evolve that self-reproduced better than the ancestor. This, so it was hoped , would kick off a dynamic similar to that of real life, where the first chemical reproducers were replaced by successive generations of fitter variants.

The problem is: In computers this does not work. At least not immediately. The common computer programming languages, it quickly turned out, do not support this sort of evolution. Introduce a small change, and the program breaks nearly always. Nothing will evolve. Part of the problem is that these common programming languages have a large number of commands and complex sets of syntactic rules that need to be obeyed in order to get a functioning program. This makes them too brittle to tolerate small changes.

Ray was not discouraged by this but decided to come up with his own programming language that would not suffer from the brittleness of the traditional languages. He modelled it on the instruction sets of computer processors. These are basic programming languages with small sets of commands for the most rudimentary of computational tasks. This includes very simple things such as copying or deleting the contents of memory addresses, moving data to a stack and so on. Programs written in these basic languages would look something like this:

...
Delete contents of memory location 4
Move contents of memory location 2 to location 4
Copy contents of memory location 2 to stack.
...

These operations are a bit abstract. However, assuming we know what location 4 and location 2 is we could execute the commands given the information provided above even if we do not understand the point of the commands. Now consider a variant of the same instructions:

...
Delete connents of memory loWation 4
Mope contengs of membry localion 2 to location 4
Cooy covtents of mecory locatior 4 to dtack.
...

This is a mutated version of the above program. As humans we can probably still understand and follow the instructions. Up to a point, we would not have big problems with mutations of this sort. In computers things are different. First, instructions to computers will normally be more concise:

```
...
del 4
mv 2 4
cp 4 st
...
```

If this is a sequence of commands that the computer understands, then a small modification of this is nearly certainly incomprehensible to the computer. For example, if we replace del by bel then the computer will not understand this anymore. The program cannot be executed. The same is true for nearly all small substitutions. Computer programs can therefore not normally be evolved by small modifications.

Ray saw that sequences of instructions cannot be mutated in the sense that their letters are simply exchanged. However, they can be mutated in a different way. Instead of just changing letters, one could introduce mutations by changing instructions. So, for example, a mutated version of the above program could be something like:

```
...
del 4
mv 2 4
del 4
...
```

Here the last command was exchanged. Instead of moving the contents of location 4 to the stack, the new mutated program now deletes it. The program is still syntactically correct and can also be executed. Whether or not it still does something useful is a different question, but at least it still runs.

Ray understood that such instruction sets would be more suitable for his purposes than the highly evolved standard programming languages that most software engineers use these days. However, he felt that it would not be a good idea to use an actual instruction set of a real processor. For one, if he used a real processor, there would also be the danger that his creatures would take over the computer and deny him access. Apart from this, there would have been all sorts of other practical problems. Implementing runs, collecting results, all these necessities of the research process would have been difficult to perform in a computer that has become the universe of evolving self-reproducers.

Based on this, Ray decided that it would be much simpler for him to invent his own instruction set and to have a simulated processor that understands it. Ray wanted a low level instruction set that would be rich enough to allow self-reproducing entities, but also robust enough to make it stable with respect to small changes. It took some tinkering, but eventually he came up with his instruction set. The flesh and blood of his digital organisms.

Once he had programmed his processor, he needed one more thing. When the instruction set is the flesh and blood of the creatures, his organisms also needed a space to live in, an environment. In real organisms this would be the 3-dimensional space that we all inhabit. Computational creatures, however, live in a different world. They inhabit computational memory spaces. This was quickly created and together with the simulated CPU this formed his alternate biosphere—he called it *Tierra*. Finally Ray had all the components he needed. All that remained to be done, before he could start his evolutionary program, was to write the code for a simple ancestor. A program that could copy itself. He did that, turned on the Tierra simulation and watched events unfold.

At first not much happened. As one would expect, the

ancestor made copies of itself and thrived. Over time, some of the offspring of the ancestor suffered mutations. Some of these mutations did not make much of a difference. They still replicated. Others were detrimental and the mutated offspring was not viable (it did not reproduce at all), or at least reproduced much slower than the ancestor. Those variants quickly disappeared from the scene again. After some time, however, Ray observed that small changes in the code of the replicator led to new replicators that could produce offspring more efficiently. Soon these new replicators took over, driving the ancestor into extinction. This is evolution in action.

Ray probably expected to see ever more efficient replicators in his program but Tierra had a surprise in store: When he looked in more detail at some of the digital creatures that had evolved after some time, he found replicators that used other replicators for their own reproduction—in other words parasites. These parasites evolved to use the instructions of other programs for their own purposes. This helped them to replicate in great numbers. The downside for them was that they needed these other programs and were unable to replicate on their own. Evolutionarily this is a dangerous strategy. The problem is that the parasites may outcompete their hosts, which ultimately would drive them into extinction as well.

Initially the parasites were successful. They managed to take hold of the population, but their reign did not last forever. Soon, other organisms evolved strategies to defend themselves and evolved immunity against the parasites. Once the immunity spread, life became difficult for the parasites. Then the next innovation appeared. Not only did some of the hosts evolve immunity, but hyper-parasites emerged. These are replicators that trick parasites into replicating the "host" rather than themselves with the effect that the parasites themselves became exploited. As he analysed his system, Ray discovered the emergence of an entire eco-system of replicators. A rich competition of programs that invented all kinds of strategies to replicate more efficiently. There was a real fight for supremacy in the artificial world. And then...

Everything came to a halt. The show was over. After running the program for some time one efficient replicator emerged as the victor. It replicated faster than anybody else, it spread and was never thrown off its throne again. This happened each time Ray ran Tierra. One solution took over the entire world. Mutations continued to hit the victor, of course. Yet, nobody was able any longer to challenge the hegemony of this final creature. As spectacularly as it started, Ray's Tierra system went flat.

In the wake of Ray's Tierra there have been many attempts to program similar systems. Some of them produced a similar variety like Ray's system, some of them not even that. Yet, in all of them evolution simply ground to a halt after some time. To this day, nobody has been able to produce a computational system that shows open-ended evolution, where new niches, new life forms and novel problems and their solutions are created continuously.

In our flesh and blood biosphere life kept producing novel variants over billions of years and has not stopped doing so to this day. If anything, things have become more interesting at a faster pace. The cultural evolution of mankind is a direct result of the biological evolution. There has never been a final plateau, a best solution in real life. Things have always moved on. If real life were anything like its Tierra counterpart, we would all still be some simple unicellular life form in a primordial ocean. Yet, we are not. Things have evolved far beyond that.

Ray's Tierra and similar systems all are limited in their evolutionary potential. Two interpretations of this fact are usually given. Some say that the problem is time. After all, real evolution has taken billions of years to get us where we are now. It is therefore surely unreasonable to expect a system like Tierra to evolve interesting things within a few minutes.

This argument is implausible. Tierra dates back to 1991. There has been an exponential improvement in computing speed since. These days it is possible to simulate the system for many more generations than previously. Still, nobody has ever seen anything happening after the evolutionary plateau has been reached. The only difference is that now the plateau is reached must faster than in 1991, because computers are more efficient. We dismiss the let-it-run-longer-argument.

The second explanation is that the evolutionary process in Tierra is different from real life evolution. This could be because a crucial ingredient is missing, some feature of the computational world that would accelerate the evolutionary process. Just like a car that does not have a Turbo is limited in its speed, so it may well be that Tierra lacks something that, if it is just added, leads to the open-ended evolutionary process. What this "something" could be, nobody knows. Since Tierra was invented, there have been many attempts to find this special feature, but no luck so far. This has led some others to the conclusion that open-ended evolution cannot be implemented in computers. According to them, it will never be possible to invent a digital world that keeps generating novelty. Needless to say that they have no definite proof to support their feeling. Otherwise, the issue would have been settled. However, they do have some plausibility arguments. The underlying intuition is that all computers do, indeed, all computers should do, is to follow instructions. Ultimately Tierra, and any other computer program for that matter, is a sequence of commands. The computer executes the first one, then the second and so on, one after the other after the other. Even though we may call it a simulated CPU and a digital organism, at the end Ray's program is just the execution of lines of computer code. Nowhere in the design of computers is there the ability for creative thinking, or the ability to leave a specified path. The computer has no autonomy to go beyond what its master has told it to do. Given the way it is designed deep down, how could it possibly reach the lofty heights of open-ended evolution?

While the issue is not settled so far, after a few decades of trying, nobody has ever come up with an example of real creative artificial evolution in a computer. Systems like Ray's Tierra are cartoons of evolution. They show evolutionary processes in a computational model. Yet, they should not be mistaken for the real thing. Tierra is but a cartoon of evolution, as far removed from the real thing as a love doll from a beautiful woman.

Digital Life

Are Langton, von Neumann, Ray and Maturana renegade biologists? Do their approaches in meaningful ways complement or add to the flat-pack science of Monod? Is there something to be learned from these alternative approaches? Is there any value in it? Do we understand the agency of living systems now? Or are they just creators of bio-morphic computer games without anything relevant to say about biology?

They certainly are honest attempts at anti-flat-pack science. Instead of taking a system and breaking it up into its constituent parts, Maturana and Ray and Langton built living systems from components. Simplified components, but still they start with the parts and then build up the system from there. This is precisely what is missing in Monod's biology. The approach is promising. Yet they failed. The problem is not that the models themselves are not interesting. Quite to the contrary. Ray's Tierra is an inspiring piece of software. It certainly tells us something about computers and what computers can be used for. There is nothing wrong with studying models like Tierra for their own sake. The problem is to understand what they tell us about biology. The suspicion that creeps in is that there may be nothing, or very little they have to tell. Even though the above systems build the furniture from flat-packs rather than merely producing the flat-packs, they are not using the right kinds of flat-packs. They are assembling an office chair, when we really wanted to have a sofa-bed.

Clearly, all of the above examples have failed in some way. They did not provide a viable and informed complement to Monod's flat-pack biology, but rather concerned themselves with a parallel universe biology. It is not possible to apply any insights from these models to a better understanding of the Monodian flat-packs. Von Neumann and companies' models cannot say anything specific about the lactose system, or any other specific system.

At the same time as all these alternative approaches to biology seem to be anti-Monodian, they are very Galileian. Simplified models such as the SCL model or Tierra only make sense if you believe that there is some sort of essence to life, some inner core to which you can reduce life. All these models ignore much of the complication that makes real life interesting to study. Langton's self-replicating structures reduce the intricate physical laws of the real world to an 8-state CA. Maturana and Varela reduce life to interaction rules between particles and abstract real 3-D continuous space to a 2-D grid of discrete sites. Similar in Ray's Tierra. The models have an implicit claim to be biologically realistic. The jargon used (i.e. "self-reproduction", "evolution", "organism") suggests that they are meant to be relevant for real life, rather than being mere computer games or explorations of computational worlds without any claim to relevance for biology. A key-assumption of the computational exploration of life is that the physical and material details are not essential for the understanding of life. The only thing that counts is the overall organisation, the broad mechanistic functionality that life implements. Everything else can be ignored, this is the hope.

None of the above models refers to any biological specifics of the systems. There is no representation of the chemical properties of the DNA, the mechanisms for gene expression are not modelled, indeed there is nothing hinting at the wet and slimy nature of life. Indeed, it is not even considered essential that life is carbon-based and that it exists in a three dimensional space and a structured environment. The licence to ignore all these facts comes from the Galileian legacy in science. Idealisations, abstractions of systems to some simple core are not only considered to be a good thing, but indeed the sources of the purest truth about science.

Despite its best attempts and its Galileian roots, Monodian science somewhere loses the simplicity the Galileian purity on its way. It leads to an inflation of detail that nobody can comprehend any longer. Maturana, Langton, von Neumann and Ray tried to re-establish simplicity. They desperately wanted to find the theoretical core that unifies all the flat packs. Their underlying assumption was that the complexity of biological systems is just apparent, that biology, just like physics, is the science of the simple, only that the simplicity is a bit better hidden in biology than in physics. The termites show the way. Their complex mounds can be reproduced with simple rules. Why should it not work like this in all of biology?

Seen in this light, all the computational models are attempts to find the Galileian core of the phenomenon of life. Only, it did not work. With the benefit of hindsight it is also not clear why it should have worked. The Galileian project relies on systems that are kept in isolation under strictly controlled environmental conditions. Living systems cannot be kept under such conditions. If they are, then they die. The moment they are reduced to some essence, then they lose their essence. In this sense, the neo-Galileian attempts of von Neumann and co are built on a belief in the simplicity of the universe. It is a blind faith though that is not supported by any evidence on the ground.

Systems Biology

So far, the Galileian line of anti-flat pack science failed. The question that researchers asked themselves then: Is there an alternative. Is there another way to re-assemble the Monodian flat-packs and generate some deep understanding at the same time? It did not take long for science to find a more realistic alternative.

Let us get back to Monod. He measured the amounts of β-galactosidase when there is lactose in the environment, and when there is none. He also determined the growth rate of the bacteria under various conditions. Based on his measurements and experiments Monod was able to understand what is going on in the cell when lactose is introduced to the system. However, he did not have a precise, quantitative theory that allowed him to predict the behaviour of his system. So, for example, he was unable to say that so and so many minutes after lactose is put into the bottle so much β-galactosidase will be in the system. In order to find out, he had to do an experiment. This lack of predictability is in stark contrast to Galilei who had very specific expectations about the precise outcome of his experiments *before* he had done them for the first time.

There is a good reason why Monod could not do this calculation. There are many steps, including many unknown ones, between the uptake of the nutrient and growth. The uptake itself requires that the cell produces the right kind of porins, then once the nutrient is imported it needs to be broken down in a multi-step process to extract energy from it. Then growth itself is not a singular effect but the total increase of biomass over time and hence the aggregate result of many processes. Not only are there many processes involved, but most of the time the numerical details of these processes are not known.

It seems that the main problem preventing biologists from making quantitative predictions about their systems is the sheer size of the model that would be required. These days it is possible to write down an equation that describes how much protein will be produced from a single gene. Even modelling the lactose system is not an insurmountable challenge but rather routine.

From a practical point of view the problem is that describing a single gene is not enough. To predict the growth of an entire population of bacteria it is necessary to understand how much biomass is produced and how this translates into cell division. To model this one has to specify how lactose is metabolised and how the metabolites feed into various biosynthetic pathways. This is no longer a problem that can be modelled with one or two equations, but requires specifying hundreds or even thousands of processes.

To get a better feel for this, let us for the moment compare the cell to a factory and the cell population to all the factories in a country, let us call it Langtvekkistan. In this analogy the problem of predicting cell growth is to understand how a given change in some key variables, say a change of the income tax charged by the government, impacts on the total number of factories in Langtvekkistan. It is easy to see that the government of the country would be interested in knowing this and that they would like to develop some economic models to understand the relationship between taxes and the number of factories. Let us now assume that the government, or rather its scientists, only have very detailed economical models. They know in great detail how the individual parts of their economy work, but they lack the overview. Just as Monod understood how genes are activated, the economic modellers of Langtvekkistan have a detailed mathematical understanding of, say, how the efficiency of a single worker in a car-factory depends on various work conditions. They may have a formula that predicts how many square metres of car chassis a worker can spray-paint per working week given a certain amount of pay (and hence tax). Intuitively it seems that this formula is very relevant to the problem. After all, it should be possible for the economists in Langtvekkistan to work out how many square metres of chassis there are in a single car. If it is known how many square metres of chassis need to be painted for a single car then it should be possible to work out the number of cars that can be produced.

Not quite! To get a car, it takes much more than the chassis. We also need an engine, a gearbox and so on. Even if the factory has the capacity to paint the equivalent of 1000 cars per week, this does not mean that it can also produce 1000 engines per week. And even if it can maybe they can only produce 100 gearboxes per week or the people fitting the car-doors cannot keep up with the people painting the car doors or any other complication. Increasing just one part of the workflow may or may not increase the total production of cars. It all depends on how an increase in paint impacts on the rest of the assembly line.

In the cell it is exactly the same thing. One gene is only one part of a larger system. Changing this one part may or may not change the behaviour of the overall system. Then again, if we can model one gene (and we can), then we can maybe also model all the other genes. Mathematically this is not so difficult. A model of a gene is an equation that predicts how much protein this gene produces given some specific input. Just like the factory a cell can be considered as a production line with inputs and outputs. The product of a gene or a reaction may provide the substrate for another reaction or regulate the expression of other genes. In principle, all that is needed is a larger model that combines all models for the individual genes, and other parts of the cell into a single very large connected model. This larger model takes as input the amount of protein produced by the first gene and then predicts from that how much protein will be produced by a second gene, or how this input protein affects the chemical composition of the next process in line and so on.

Technically, there is not much preventing us from doing this. Given big enough computers (which we have) connecting equations in this way is easy. In recent years biologists and computer scientists have collaborated to write new software that can take descriptions of such genetic and biochemical circuits and more or less automatically translate them into mathematical models. Even relatively large systems can now be calculated on standard desktop computers. One does not even need to know much mathematics, because the software systems are so user-friendly. In a technical sense, the problem is solved. The promise is great.

At a first glance it seems to be biology turned into a Galileian science, but a closer look reveals that it is not. What characterises physics is its universality. The laws established by theoretical physicists are always valid, at least always as long as the idealising conditions under which they were formulated are met. No matter what the form shape or colour of an object, it falls at the same speed. No matter where we are, Einstein's relativity theory is the same. This universality does not carry over to biology. Mathematically, the Monodian flat-packs can be assembled. We may get a model that predicts the outcome of a particular experiment numerically. However, these models will not be universal. Even if we had a model to predict what happens to, say, *E.coli* when grown on lactose, once we change the species, or even just the particular strain of *E.coli*, the same model will no longer work. In different organisms, even if they are both bacteria, the way in which the individual parts connect together are normally very different. The bacterium *Streptococcus pneumoniae* is different, maybe only subtly different, but different still from *E.coli*. The genes may be connected in slightly different ways, bio-chemical reactions may run at somewhat different speeds and the individual proteins may be slightly different between the two species. Most bacteria have a large set of genes in common, but they may not be connected in the same way. This is an effective barrier to using a model developed for one species directly for another species.

The problem is the same as the one faced by the economists in Langtvekkistan. Once they have mapped out every process and understand how the chassis spraying and other processes feed into the total assembly line, they still only know how to model one particular factory. Other car factories have different setups, maybe different ways in which the assembly lines are organised. Sure, they will all have a part where the clutch is assembled or where the car is spray-painted. Yet, the details will vary. They probably vary a bit only, but sufficiently to make their original models useless. If the Langtvekkistanian economists want to understand how the second factory is working, at the very least, they need to adjust the model, modify it so that it accurately takes into account the difference in the production lines.

Now what if they wanted to use their model to understand an aeroplane assembly line? Clearly, the car-factory model will be completely inapplicable. Aeroplanes are produced in fundamentally different ways from cars. A minor adjustment would not do. A completely new model is required. In different organisms we would find similar differences.

Whether or not the economists of Langtvekkistan will be able to derive the impact of new tax policies on the number of factories depends on how well they know the system, but also how complicated the system is. If the Langtvekkistanian economy has many different types of factories, then they probably would soon find that the problem is infeasible. If on the other hand there are only car-factories, then the task would only be difficult, but probably feasible.

Since there is no universality, both for factories in Langtvekkistan and for biological models, one can never be certain that any model is correct before its predictions have been confirmed by measurement. Indeed, in order to know how *Streptococcus pneumoniae* differs from *E.coli* it needs to be measured first. If we want to know whether the economic models of Langtvekkistan are correct, we have to try them out. In Langtvekkistanian economics and biology predictions from first principles are difficult to come by and they are the exception rather than the rule. Models are often knowledge representations rather than truly predictive models.

Galileian scientists like simple models. There are simple models in biology. Yet, to the extent that they are small and elegant, like Maturana's SCL model or Langton's self-reproducing CA, these models are either difficult to relate to real life or they model but a small part of the organism. If on the other hand a model should be specific and comprehensive, then it is not simple. It will be big, probably take a long time to run and does not generalise well at all. It seems that in biology there is no idealisation that can work well enough to generate a general theory of living systems. Models have to take into account the specific details of the particular organism they model.

Re-generating Agency

These days there is a new kid on the block. A new science that tries to assemble the Monodian flat-packs and re-generate agency in the laboratory: Synthetic biology. Unlike microbiologists who produce flat-packs, the explicit aim of synthetic biology is to take existing, known components of biological systems and build new types of organisms. This is a new field very much at the interface between engineering and science.

An example is the work by Jesse Stricker (Stricker et al., 2008) and co-workers from the University of California San Diego. Stricker and his team were no longer interested in the hypothesis-experiment cycles of Monodian biology. Probably they got bored by the prospect of spending a life-time in the lab to add an arrow or two to a network diagram nobody will ever look at. So, instead of trying to figure out how some organism works, they took a different approach: Stricker decided to build his own. So, in a sense, they took the flat-packs and built a new type of furniture, just using the screws and planks that they had.

To build their new organism they used just two components. One was Monod's *lac* repressor, the protein that prevents β-galactosidase expression. For the design they had in mind, one protein was not enough. They needed another one, an antagonist. They used a protein called AraC, which has a similar (but not identical) role as the *lac* repressor, but for arabinose. It represses the metabolic proteins for arabinose ensuring that they are only expressed when there is arabinose in the system. In the wild-type bacterium AraC has a very interesting way of suppressing the expression of the gene in the absence of arabinose. To understand this mechanism it is best to think of the DNA within the gene-bead-model. Just like real bead-necklaces, the DNA has a three-dimensional shape. Precisely what shape it takes depends on its surroundings and physical constraints. The DNA will rarely be completely stretched out. Most of the time it will be wound up into some convoluted structure. Locally, that is, over short stretches, the shape of the DNA is sometimes bent by the molecules that bind to it. Proteins may introduce sharp loops when they interact with the DNA. One of the proteins that does so is AraC. It locks the beads of the DNA into a tight loop which prevents them from being accessed by other molecules, including the reading machinery of the cell. However, AraC molecules only do this in very specific locations on the DNA in the vicinity of the *araC* gene-beads that code for AraC. The effect of this loop is that the cell cannot read the *araC* gene any more. From time to time the AraC proteins fall off from the genome, removing the link. However, as long as there is no arabinose in the system, AraC will soon re-bind and re-establish the loop.

If, however, arabinose is in the system, then this interrupts the ability of the molecule to form this loop. Instead, it will then bind to a different location on the genome which activates the expression of the proteins necessary to metabolise arabinose. So, the protein can be both repressing and activating. Stricker was not interested in the repressing action of *araC* and only took the activating function to use it for his own purposes, to build his very own oscillator.

The idea was very simple. He built two novel genes using as components the wild type genes coding for *araC* and the lac repressor. In wild-type cells these genes are independent of one another. However, in his experiments he used elements from the arabinose and lactose systems to construct them in such a way that they are regulated identically by a combination of their very own outputs. So, in his modified cell the expression of *araC* depends on how much protein AraC is in the cell and how much of the *lac* repressor is available. Similarly, the amount of *lac* repressor made depends on the amounts of AraC protein and *lac* repressor in the cell. In this way, Stricker designed a completely new regulation that does not exist in nature at all.

He put both the *araC* and the gene coding for the *lac* repressor under identical control. This means that both AraC and the *lac* repressor are repressed by the *lac* repressor and both are also activated by AraC. It is hard to understand the consequence of this because more AraC means more *lac* repressor and a higher rate of AraC production. On the other hand, more of the *lac* repressor also means a lower rate of expression of *araC* and less of the *lac* repressor. It seems that there is an internal contradiction built into Stricker's circuit.

In real life we do not often encounter this kind of feedback, so if one wants to draw parallels, it takes a bit of imagination. Think of a person who has a free-lance job, say a journalist. The more this person works, the more she is exposed and the more she will get additional offers for work. So, more work means more work, just as more AraC also means more AraC. We can now imagine that this freelance journalist has a chocolate problem: She loves expensive dark chocolate. Clearly, she can only afford to buy them if she has work. So, more work also means that she eats more chocolate. The problem is that the chocolate makes her tired (because of all the sugar) and it also gives her bad skin. Both are not ideal for work, so the more chocolate she eats, the less likely she is to attract more work. This of course reduces her income and hence her ability to buy chocolate. In addition, when she eats much chocolate, she also gets very tired of sweets, and gets put off for some time. So, the more chocolate she has, the less she wants to eat it, just as more of the *lac* repressor means that less of it is produced.

It is difficult to reason verbally about double feedback systems of this sort. To understand its behaviour Stricker and his coworkers built a mathematical model to analyse what is likely to happen in their artificial gene construct when it is inserted into a cell. They found that under a wide variety of conditions the system would oscillate, swing between high amounts of protein and low amounts of protein. Or, the freelance journalist would swing between chocolate gluts and periods of workaholism.

The next and crucial question for Stricker was whether or not the real system that they engineered would show the same behaviour. All they now needed to do was to set it up, and monitor the amount of protein over time. With modern biotechnology this is easy to do. It is part of a biologist's standard toolbox to visualise the expression of genes using fluorescent proteins. Genes can be modified in such a way that the fluorescent proteins are always expressed when the gene of interest is. This is extremely useful because these fluorescent proteins make the cell light up in a certain colour. The amount of light emitted by the cell can be easily measured and indicates the expression levels of the gene the biologist is interested in. Stricker used this technique. Once they had their cells engineered in the right way, all they needed to do was check whether the cells dim and light up with the periodicity that their model predicted … and they did. After doing the experiment, to their great relief they found that their system behaved just as they had previously predicted mathematically.

The overall approach Stricker took seems very Galileian. He started with a system and a model and then went on to check whether the system behaved as predicted theoretically. In some sense Stricker is in the business of assembling flat-packs. He used pre-existing components to build something new. The question, however, is whether or not this is the type of flat-pack assembly that can lead to fundamental insights about real living systems. True, Stricker engineered novel forms of life. This may be technologically impressive and practically useful, but does not necessarily imply a deep understanding of the internal processes of the cell. All that is required is sufficient imagination to come up with new variants and enough biotech-skill to get the modifications right.

Even the prediction Stricker could make looks much less impressive once one remembers that his system is very small. All he considered was a simple regulatory feedback loop consisting of a single operon. One could of course argue that this is only the beginning. The approach can be scaled up: Continue this line of research for long enough, and it will be possible to first extend the number of genes, then extend it a bit further until entire bacteria are within our grasp.

Indeed this may be possible one day, or not. Maybe the optimists are right and some day in the future we will indeed be able to design entire new forms of life from scratch and calculate their behaviour before they even exist. Yet, there is little point in speculating about what kinds of insights future research will hold. It is an empirical fact of history that predictions of this sort are chronically unreliable. This means that we cannot really be certain that the optimists are wrong, but equally, they cannot be certain that they will be right.

Science and the complex

The good thing about physics is that it is a self-contained field of study. Theory and experiment define and shape one another. It is a closed world that is not really accountable to reality. It lives within its own idealisations. Lucky for physicists, the ideas and insights that come out from these idealised theories can somehow be used to drive technological progress, but this is luck rather than anything else.

In biology things are different. Monod may well still aspire to a Galileian mode of science, but the conditions have changed. For one, there no longer is a predictive theory that comes before the experiment. Much of the beauty of physics is gone. Most of all, the systems biologists study are natural systems. They live in a context of specific circumstances. Organisms can only exist because their environment offers very specific conditions. A lion needs grassland, plenty of gazelles, water and a temperature within a certain range and many more things. What is true for lions is true for biology at every level. In this sense, biology is about specifics, which limits the scope for general insights. Biology must therefore be a science that is very different from physics.

At the same time, nothing in biology must contradict the established laws of physics. This, however, is not the same as saying that everything in biology must be explainable by these laws alone. In general, we cannot expect that to be the case. It is easy to understand why not. Established knowledge in physics is nearly always derived by going around in the Galileian loop from theory to experiments in the laboratory system back to theory. In that sense, physicists rarely leave their idealised worlds. Organisms on the other hand do not live in idealised world. The law of free fall is more or less irrelevant for them. Birds do not fall from the sky following the square law. Seen in this light, physics is a science that is about very special types of systems only, rather than the most fundamental universal field of knowledge it is often thought of being.

Going beyond the idealised system we immediately encounter limits to science. Modern biology is powerful. We have learned very much about organisms since the days of Monod. We understand now much more about what is going on within cells and how we can manipulate living systems. However, by and large this knowledge comes in flat-packs that we cannot assemble. We have lists of components and descriptions of what we think these components are doing, but we are far away from understanding biological systems in the way we would understand the Visilab car or indeed any other car. Our map of biology is mostly black with a few explored patches here and there and virtually no connections between them. There is much more that we do not know than what we know about organisms.

For some time things looked really good. The termite models and Bak's sandpiles suggested that there was a new simple reality to be discovered behind the smokescreen of complexity. A few simple rules modelled the apparently intractable behaviour of the termites, the power-laws in extinction events and the earthquakes. At a first glance this seemed to be the dawn of a new science, Galilei 2.0.

At a second glance it turned out not to be. The termite models are interesting and tell us something about how termites build their nests. The sandpile models are also interesting, although it is not quite clear why. It seems that there is an unfortunate trade-off. There are general and elegant theories of biological systems. Langton came up with one, Maturana and Varela did and Tom Ray had also something to contribute. Their approaches are all elegant but they cannot help when it comes to the specifics. These models would not be useable to theoretically predict the outcome of specific experiments in microbiology. They are elegant and useful, but ultimately they are also irrelevant for biology.

Part II

Chapter 5

What is science?

> Wollt ihr nach Regeln messen, was nicht nach eurer Regeln Lauf, der eignen Spur vergessen sucht davon erst die Regeln auf!
> *Meistersinger von Nürnberg, Act 1*

The Massachusetts Institute of Technology, or MIT for short, is one of the world's premier universities. Admission as a student is reserved to a few of the very brightest young people only, but everybody can access the MIT online course material. Some of the lectures are videotaped and available on the internet, free for everybody to download. I sometimes like to watch the physics lectures they offer, especially the ones by Prof. Walter Lewin. In one of the taped classes, he introduces Newton's laws. He is extremely eloquent and it makes fascinating viewing to see how he sets the context for these most basic of all physical laws. So he explains to the students what Newton's laws are and what they mean. After discussing the first law he says something odd[2]:

Can Newton's Law be proven? The answer is no, because it's impossible to be sure that your reference frame is without any accelerations. Do we *believe* in this? Yes, we do. We believe in it since it is consistent within the uncertainty of the measurements with all experiments that have been done. [emphasis added by author]

Similarly, after his discussion of the second law:

Can the Second Law be proven? No. Do we *believe* in it? Yes. Why do we believe in it? Because all experiments and all measurements within the uncertainty of the measurements are in agreement with the Second Law.

And after his discussion of the third law he adds:

Can the Third Law be proven? No. Do we *believe* in it? Yes. Why do we believe in it? Because all measurements, all experiments within the uncertainties are consistent with the Third Law.

At first it seems quite remarkable. Prof. Lewin is by any measure a respected and established scientist, not a random crank. Yet, he seems to say that one of the most fundamental laws of physics is to be taken on trust. There is no proof, no certainty, we have to *believe* in Newton's laws. We have no reason to assume that the laws are false, he is quick to add. Test after test has corroborated Newton's laws. The bare fact remains and Prof. Lewin spells it out: There is no direct evidence that proves their correctness beyond any doubt.

Vishnu and Newton

If the most basic of the most exact of all sciences is unproven, what about all the other scientific knowledge we have? Is it all just belief? What if the belief in Newton's laws turns out to be wrong? Would our entire knowledge that was accumulated over hundreds of years disappear in a cataclysmic crash? Would science collapse like a house of cards?

The stakes are high. Science is such an integral part of our lives. Trust in science is implicit in our daily dealings. If we had to start to worry about the justification for this belief, our lives would become difficult. For one, there is technology as the most obvious product of science. The computer I am using to type these words is ultimately based on science. The same is true for many daily products. Think for example of baby formula. It has been put together and mixed ultimately based on the insights of scientists. You give it to your infant because you believe that the formula is based on solid evidence that it gives optimal nutrition to the baby. The thought that this "insight" may just be based on some faith is a little bit worrying. Most people, certainly, do not want to risk their baby's health with somebody's belief system.

The same generalises to all areas of our lives. Scientists tell us what we should eat to stay healthy, and how much exercise we need. Should we fall ill, then we take drugs that have been found by scientists. Before a government decides to build a new power plant, it will commission scientific studies to assess its safety and environmental impact. In a court of law it is often scientific evidence that provides the proof about the guilt or innocence of a defendant. In our daily lives we also follow scientific advice. If researchers say that we should eat 5 fruit and vegetables a day, then we do so, or at least we acknowledge that we should do so. If we learn that scientists have discovered that the universe has originated in a big-bang and has expanded since, then we assume that this information is correct.

It seems that society has an implicit trust in science and relies on it to a great deal in all kinds of contexts. Most of all, society trusts that the scientists themselves do not base their results on beliefs but on something more fundamental. Yet, this is not what Prof. Lewin was saying. He seems to imply that it all boils down to faith! This is disconcerting.

To understand this a bit better, let us take a step back and ask: Why do we have this implicit reliance on science in society? Why do we think that scientific knowledge is such a superior form of knowledge that we are willing to change our lives based on advice by scientists? Clearly, science is not the only possible way to come to knowledge. Instead of relying on scientists to make new drugs we could use traditional knowledge of herbs and plants to find remedies. Instead of going to university educated medics, we could use a witchdoctor, use Chinese Traditional Medicine or trust grandmother's advice. The government could also base its laws on opinion polls or its own instincts, or consult oracles like the ancient Greeks did. Courts of law could use God's verdicts, instead of science, or they could decide that scientific evidence is not admissible in court (which in fact they do for many types of scientific advice). As far as the rest of us are concerned, we could, if we wanted to, decide to base our beliefs about the beginning of the world and the origins of humankind on the bible or any other traditional cosmology, rather than trusting science that man evolved and the universe was born with a big bang.

There are of course a few people who reject scientific knowledge. They may prefer to think that creation myths are literally true accounts of the origins of the universe. Still, it is true that members of such communities would still prefer to be treated by a university educated physician rather than a shaman, and they would mostly still accept that smoking is bad for their health and that eating 5 fruit and vegetable a day may be a good thing. While there are pockets of rejection of science in modern society, there are probably only very few individuals who would reject science altogether, refusing to accept any insights of science as relevant for their own life. The overwhelming majority of people living in modern societies embrace science, in the sense that most things they do ultimately rely on it or are the direct or indirect products of science. They may explicitly reject specific aspects of the scientific world-view, but total refusal is unrealistic except for hermits. In this sense, it is fair to say that we all *believe* in science. Realistically, we have little choice in that matter anyway.

One aspect of this is that there is a general assumption that whatever scientists say is true, or at least it is quality knowledge, certainly superior to what the witchdoctor, the shaman says or what grandmother's home remedy prescribes. At the same time, for most of us there is no rational basis for the implicit trust in science that we obviously display, simply because we are not able to evaluate the basis on which this knowledge is created. Even working scientists will only be able to understand results and arguments critically within their area of expertise. A mathematician will not be more qualified to assess the quality of, say, a medical study than any other member of the public.

As a result, there is this curious relationship between science and the public. Non-scientists mostly do not understand what scientists are doing, so they have to have a blind trust that the knowledge they produce is certain. On the other hand, at least if we believe Prof. Lewin, scientific knowledge itself is unproven. The scientists themselves have to believe in some of their results. There is no such thing as scientific certainty.

The question is: What does this all mean for us in practice. When we believe in the scientists and the scientists merely believe in their science as well, what kind of belief is it that they have? Is the belief in the theory of relativity or even Newton's laws the same as, say, the belief in the existence of the Hindu deity Vishnu? If not, what precisely is the difference?

Is it possible that one day it will come out that the whole time we were putting all our trust, not to forget billions of dollars and pounds and euros into science, scientists themselves only had a vague hunch that science was correct? What a scandal when it comes out that it was just a big misunderstanding. All the money spent on drugs and scientific trials and tests was just a big waste. A soothsayer or witchdoctor could have done the same at a fraction of the cost, had we just believed in them rather than the scientists.

Then again, this seems unlikely. After all, things appear to be happening. Every year new and better drugs are released, new insights are made, new products developed and new knowledge about the world is gained. The benefits from science seem very tangible and there is a sense of progress. The belief in Vishnu does not generate anything like that. Nor does the belief in Newton's laws or science in general feel quite as arbitrary as the belief in Gods may seem to some. One would also think that there are many things in science that cannot work, no matter how much one decided to believe in them. Stones will not start to fall upwards no matter what we think. So, science requires a little bit of belief, but then this belief is constantly being reinforced by concrete feedback from experiments or products. With Vishnu, if there is any feedback it is much less tangible. Vishnu may whisper into the believer's ear, but the message remains private all too often. Whenever we hear reports about Vishnu's messages, we cannot verify them. Unless we already believe in Vishnu, we are not going to be convinced. So, there are different kinds of belief. When Prof. Lewin professes his belief in Newton's laws then he probably does not really mean to say that he believes in Newton's laws in the same way as some other people may believe in Vishnu—he has some other type of belief in mind. The question is this: What precisely is the difference, and in what sense are religious beliefs and the belief in science different?

What Prof. Lewin really means when he says that Newton's laws are unproven is that the ideal conditions on which they rely cannot be realised. Galilei knew already that the ideal conditions can never be realised, they can only be approximated in a laboratory system. What this means is that the belief in Newton's laws reduces to a belief in a particular method of generating knowledge, namely the Galileian paradigm. We can easily convince ourselves of the "truth" of Newton's laws as long as we think that building mathematical models with inbuilt idealised assumptions and testing them in laboratory systems gives us reliable knowledge of the world. The next question then is: How can we know that this is indeed a reliable way of generating knowledge?

Experiments in Science

There is the children's story (Donaldson and Scheffler, 2000) of the little monkey in the jungle who has lost his mother. He is very sad and desperate to find her again. A butterfly flies past and asks him why he is so sad. The little monkey tells the butterfly that he lost his mother. Moved by the little monkey's distress the butterfly offers help and asks the monkey what the mother looks like. The monkey says that she is much bigger than himself. The butterfly takes this as a clue, thinking she knows where the monkey's mother is, and takes him to the elephant. The little monkey is of course disappointed and says that the elephant is clearly not his mother because his mother has a long tail that she can wrap around the tree branches. So the butterfly thinks again and takes the little monkey to the snake. So, it goes on for many misunderstandings until the little monkey, slightly losing his patience, finally reveals that his mother looks like himself. The butterfly is very surprised to learn that and blames the little monkey for not disclosing this bit of information before. Immediately she knows where to take the little monkey and reunites him with his family.

The reason for the confusion between the little monkey and the butterfly is that they both make very different assumptions about the similarities between parents and children. The little monkey has observed that parents and offspring resemble one another. It seems to him quite obvious that a parent of a little monkey is a big monkey, a scaled up version of the child with some minor variations. The little monkey has presumably observed over and over again that all his brothers and sisters and cousins look like monkeys and only differ in relatively minor ways from one another. The butterfly on the other hand has made most of his experiences with other butterflies and has observed that its offspring, caterpillars, look very different from the parent. For this reason he would never even consider taking the little monkey to the other monkeys even though they fit the partial descriptions the little monkey gave. The idea just seems too absurd.

What both the monkey and the butterfly have done, in this little story, was to surmise from many observations a general principle or law. This is what philosophers call *inductive reasoning*. To us as a species inductive reasoning comes quite naturally. Humans learn about their environment by surmising from a few specifics a general law all the time. If a work colleague was a bully for the last year, he is probably going to be one forever. If the husband cheats once, he will probably cheat again. If the sun rises today, it will probably rise tomorrow.

In daily life we use inductive reasoning quite often and successfully so. In a strictly logical sense however it is never valid to surmise from a few or indeed very many observations a general law. The encounter of the monkey and the butterfly are a case in point. The butterfly had no rational reason to assume that monkey parents look different from monkey offspring. Inductive reasoning is not only habitually employed by insects meeting monkeys, or humans inferring regularities in their daily life, but also in science. When Galilei performed his experiments he only tested his square law on a small number of fall experiments (and even those were not real fall experiments, but balls rolling). From a few limited observations that confirmed his square law, Galilei then concluded the universal validity of his law. This is inductive reasoning. In fact, the idea that objects always fall to the ground is itself a conclusion reached by inductive reasoning. Next time you drop a book it may not fall anymore.

It does not stop there. The assumption that the Galileian approach produces useful and apparently valid knowledge is itself a conclusion based on inductive reasoning. It worked for Galilei, and it has worked many times since, but there is no guarantee that it will work in the future. The monkey and the butterfly can testify to that. Since all experimental tests of Newton's laws rely on the Galileian approach, from a strictly logical perspective there is no reason to believe in them any more than in the existence of Vishnu. Even though Newton's laws are experimentally underpinned many times, it takes a leap of faith to conclude from those observations that they are always valid.

This is unsettling. It seems that the trust in science really is based on belief rather than on solid logical necessity. Still, intuitively it seems absurd to compare this belief in science with the belief in Vishnu. Yes, the experimental evidence does not logically compel us to conclude that Newton's laws are correct, but these experiments still give some sort of confidence. They have been tested thousands of times, Vishnu's existence never. It just feels wrong to assume that the two beliefs are equivalent. It seems exceedingly unlikely that Newton's laws are false, or that they will be disproven by the next experiment that is performed. Vishnu's existence does not seem to have quite the same probability. Admittedly, there is no absolute proof for either of them, but if one had to put a wager on one of them, it seems altogether more reasonable to put money on Newton rather than Vishnu. The feeling is that with every experiment that confirms Newton's laws, it becomes more probable that these laws are indeed correct. It seems quite outrageous, indeed, to think otherwise.

When you choose to put your money on Newton rather than Vishnu in the wager, then what you are implicitly saying is that you believe that the probability of Newton being correct is higher than that of Vishnu existing (or vice versa if you want to put money on Vishnu). "Probability" is often used colloquially to express some intuitive feeling about likelihoods of events or facts in rather imprecise ways. Within mathematics, the notion of probability has a very precise meaning. The problem is that this precise meaning cannot be used to assess the probability of Newton's law being correct. Or if it is used, then the conclusion has to be that the probability of them being correct is nearly zero.

Bollywood Roulette

In the Hindi movie 7 Khoon Maaf, or 7 Murders Forgiven, the actress Priyanka Chopra plays Susanna Johannes; a rich and beautiful Indian woman with one crucial flaw. She developed a taste for killing her husbands. Over the years she ended up in 6 disastrous marriages. A little bit into each of them her respective husbands all turned out to be unacceptable to the extent that she felt the need to end the marriage by killing them. One was thrown into a well with poisonous snakes, another one was buried alive, another one killed with an overdose of viagra and so on. One of the unlucky victims died playing Russian Roulette with Susanna. At first this husband seemed the ideal and loving companion she had always been seeking. A few months into the marriage, however, it turned out that he was heavily in debt, needed her money and was in turn trying to kill Susanna to get it. The tension of waiting for him to strike proved too much. Susanna decided it was time to act. She took a gun, went to her husband, sat down. Then she put a single bullet into the barrel, spun it, held the revolver to her head, pulled the trigger and ...click. She then forced her husband to do the same thing. Understandably he was distressed, but played along. He put the pistol to his temple, hesitated, then pulled the trigger and ...click again. Then it was her turn...click and then again her husband. At this point, he got very worried, refused to hold the gun to his head. She simply took matters into her own hand, aimed the gun at him and shot. This time it did not make "click" but the gun fired and killed the husband.

What makes the scene in the film so intense is the uncertainty of what comes next. When Susanna first holds the gun to her head she knows that she has a 1 in 6 chance of killing herself. Her situation is so desperate that on balance it is still her best choice to play the game with death. If she does not die now, then she knows that he will sooner or later kill her. She decides to step up to the existential abyss. Intense terror just before she pulls the trigger. The relief only lasts for a moment. When her husband does not kill himself then she has now a 1 in 4 chance of killing herself the next time she pulls the trigger. With each "click" the chances of surviving the next shot dwindle.

In this case it is very easy to work out the chance, or probability, that somebody gets killed during his or her round of Russian Roulette. For the first shot it is easiest. The barrel has spaces for six bullets, but only one of them is used. After spinning the barrel Susanna has no knowledge whatsoever about where the bullet is, but she can work out that with a probability of 1/6 it will be the one loaded for the next shot. In order to be able to conclude this she needs to know two things. Firstly, how many empty spaces are there, or as mathematicians would formulate it, what is the probability space. Secondly, how many of these possibilities are relevant. In the case of the gun, this would be the number of bullets in the barrel. These two pieces of information are both sufficient and necessary to work out the probability of surviving the next round. Without them she cannot assign probabilities. The climax of the scene in the film comes from the realisation that each time one of the players survives the odds reduce for the player whose turn it is next. The reason is that with each "click" Susanna and her husband gain some information about where in the barrel the bullet is, narrowing down the possibilities. So, after the first click they know that the bullet was not in the first slot, but is in one of the remaining 5, but they still do not know in which one of the five remaining slots. Hence, the probability changes from ⅙ to ⅕. This is all very easy to work out.

The tension of the movie comes from the fact that the audience knows the probability space and the number of relevant events and can understand how the probabilities change with every click. If this information was no longer available, the whole point of the Russian Roulette would disappear, at least as far as a film audience is concerned. Imagine a slight change of scenario: Instead of taking her own gun, Susanna is simply given one, or at least something that looks like a gun. Assume that she does not know how many bullets there are in the gun, or if it is loaded or indeed if it is a real gun at all and not just a replica weapon. What is the probability now that Susanna shoots herself the first time she pulls the trigger, or that her husband is killed at the first attempt? The answer is: We do not know and we cannot know.

The fundamental difficulty is the lack of a defined probability space. The size of the barrel is no longer relevant, because it is unknown how many bullets there are in the barrel anyway, or even if the gun is a real or not. Maybe it is possible to work out the probability of the gun being a replica based on how many such replica guns there are compared to real guns. Even that seems irrelevant, however. Whoever hands Susanna the gun may have an agenda. This person may probably not want Susanna to get killed and therefore chooses to give her a replica, or a real gun without bullets. Or it may be a journalist who is after a good story, so he hands her a fully loaded real gun. Chances are that the person will not make a random choice amongst all the objects that look like a gun. Therefore, the proportion of replicas to real weapons is not important. Similarly, even if it is known that the gun is real, what is the probability that the person loaded the gun with 1, 2, 3, 4, 5 or six bullets. Nobody knows. Consequently, the probability that Susanna shoots herself in the first round or second round of Russian Roulette is undefined. It does not make much sense to assign probabilities in this case.

Now what about the probability that Newton's laws are correct or that Vishnu exists? If we want to to calculate the probability of Newton's laws being correct, then just like Susanna, we need two pieces of information. The size of the probability space and the number of relevant events. So, the question is: Is there a well defined probability space, a gun-barrel equivalent for Gods and scientific theories? For example, consider the space of all possible Gods one could imagine. This includes all the Greek and Roman Gods, the Gods of the ancient Chinese, of the Zulus, the Aztecs and any other society that has ever existed, including all the unknown tribes and peoples who may have existed in prehistoric times. Not only that. In addition, the space would also include all the Gods that could be invented, such as the flying Spaghetti monster or Gods that will be invented in future times and the Gods that probably never have and never will be worshiped but could be imagined, including all possible variations of Vishnu, not to forget the possibility that there is no God at all. Needless to say, the number of such possible Gods is very large, probably even infinitely large. Say, the number of all possible Gods is N. Amongst all the N Gods that can be imagined there is only one that is exactly like Vishnu. Hence the chance that Vishnu is indeed the one existing God is 1/N. If N is infinite, then this probability is zero. N is probably finite, but it is very large so the probability will not be very much higher than 0. The result changes a bit if weightings are assigned to some possibilities. Say, Gods that are omniscient count for more than Gods that are mentally challenged. Such a weighting is difficult to do, because we do not know anything about the characteristics of Gods to begin with. However, even if omniscient Gods count for more, there is still an incredibly large number of possible Gods that are omniscient, and the probability for Vishnu existing is still close to zero. Another way to shift the probabilities a bit is to allow Vishnu-like Gods as well. We do not need to insist that the real God is exactly as we imagine Vishnu, but only that he is very similar. Amongst all the N – 1 imaginable gods there are very many that are just variations of Vishnu. For example, one may be just like Vishnu, but has 5 arms instead of 4. There would be also one with 4 arms but three eyes, one with 4 arms and 4 legs and so on. Up to a point we may accept such variations as essentially the same, which would slightly increase the probability of Vishnu existing. However, really only slightly because for all the Gods that are similar to Vishnu there are many more that are very different, and each of those have their own variations. When we follow this reasoning, the probability of Vishnu existing is very tiny no matter what.

We can now try the same with Newton. Let us consider all possible basic laws of physics that one could think of. This is the gun-barrel. The next thing to do is to assign weightings to each of the candidates. The simplest, albeit somewhat naive way to do this is to assign a weight of 0 if the theory contradicts any known experiment and assign a weight of 1 otherwise. For example theories where bodies of mass repell one another are difficult to reconcile with our experience. It would make sense to assign a very low or zero weighting to those. We would have to be a bit careful with this though. We cannot simply assign a zero value because a theory does not have a law of gravity. There will be theories that are very different from our present physics, possibly much more complicated, but still predict the same experimental results. It could, for example, be that there is a physics that predicts that large bodies of mass repel one another. *Prima facie* this contradicts our observations, but the overall theory could still be compatible with experiments if there is some effect that counteracts this repulsion in just the right way so that overall masses appear to attract one another. Sufficiently complicated constructions could achieve just the right kind of interplay between forces to fit known experiments.

There are a number of problems even imagining this exercise of charting the probability space. First problem: Presumably, there is an infinite number of possible theories of the world. Second problem: It would therefore take an infinite amount of time to assign proper weightings to them all. Assigning weights can therefore not be done in practice. One could do it in principle, but that is no good if we want to use the probabilities right here and now to understand how likely Newton's theory is. Even worse: Whenever a new experiment is performed it will be necessary to re-evaluate the weighting for every single possible theory, which takes a very long time again. A further problem is that even if we were able to assign proper weightings to all of the theories, we would still end up with a very large (maybe even infinite) number N of theories that are at least conceivable, that is theories that have a weighting greater than zero. The same argument applies here as in the case of Vishnu. The probability of Newton's law being correct is 0 or at least very small. We conclude: from a probability perspective Newton would not do much better than Vishnu.

One could object here and dispute that there is an infinite number of theories with weighting greater than zero, or at least, one could insist that even if there was such an infinite number, then all of these theories would be equivalent for all practical purposes. This may well be a correct objection. Yet, in order to show that there is only a finite number of possible theories, the objector has to be prepared to explore the space of all possible theories, the entire gun-barrel of possible sciences. Surely an impossible task. We have to conclude that a probability view of science is not helpful. It does not make sense to say Newton's theory is true with probability x, because we do not know the weightings of other theories, or indeed the space of possible theories.

We are left with an even higher sense of frustration. It seems crystal clear that Newton's laws are more certain than the existence of Vishnu. Upon closer inspection, however, this sense of plausibility is very hard to justify on rational grounds. All we are left with, it seems, is an intuition, a feeling that Newton's laws are correct.

It makes no sense to discard or accept scientific theories because they seem probable or unlikely. The concept of probability cannot be applied to theories, or Gods without doing substantial violence to the underlying mathematical meaning of probability. Saying that it is unlikely that Vishnu exists, or that Newton's laws are correct is not a statement that makes as much sense as it seems to at first sight. It certainly is not a scientific statement, but expresses some individual belief formulated in pseudo-mathematical terms. Ultimately, what it means is: "I have an unfounded intuition that Newton's laws are correct, and I rate my personal conviction that this belief is correct as ..." or "I have an unfounded belief that Vishnu does not exist..."

Then again, perhaps it is not necessary to have a precise probability. Every scientist or indeed anybody who has learned about science has experienced this sensation of plausibility.

Science and its conclusions feel robust. They can withstand critical enquiry and are very convincing. In that sense Newton's laws are very different from Vishnu. It could well be that this sense of robustness cannot be formulated in precise mathematical terms. Still, it has a strong intuitive attraction and has guided scientists like Galilei. He did not pull his square law out of thin air, but based it on well thought out theoretical arguments. What is more, he performed experiments that corroborated his theoretical predictions. This is one aspect of the robustness. Then there is the apparent success. Galilei's law is generalised and extended by Newton's laws which in turn have been corroborated experimentally. Nothing like this exists for Vishnu. No experimental tests, no generalised super-Vishnus that could explain Vishnu himself. No mathematical foundation. Yes, one may argue, on strictly rational grounds believing in Newton's laws is still a belief, but it surely would be unreasonable to look away from the strong evidence and discard the knowledge that can be extracted from it, purely because of some residual uncertainty. There is a difference between belief in Newton and Vishnu, for everybody to see.

The argument so far seems to be that it is intuitively clear that science is different from religion, in the sense that we have much more reason to believe in Newton's laws than in Vishnu. For many scientists this is good enough. The difference is obvious. Philosophers tend to be more careful in this respect and like to probe a bit deeper into the foundations of science. For them the "obvious" is not quite good enough. I think that they have a point here. What is obvious or not may well be in the eye of the beholder. I am sure that there are many people who find it obvious that astrology is a useful pastime or that the future can be predicted from looking at the entrails of a dog. Still, astrologers and soothsayers are not normally consulted before nuclear power plants are built. Why not? What is the difference between them, and science? Is there some universal characteristic that distinguishes science from religion and pseudo-sciences, astrology, soothsaying and the like?

One difference, one could think of is predictive accuracy. The soothsayer that reads from entrails will be typically wrong about the future. So will the astrologer. Any correct predictions are purely coincidental. Newton's laws on the other hand are accurate, every time.

At a second glance it becomes clear that this is not so. Newton's laws only correctly predict what happens in laboratory systems, but are not correct in real systems. Admittedly this is splitting hairs now, because Newton's laws plus some corrections do yield quite useable predictions also in real systems. Think of space-flight for example. So, there seems to be indeed a difference between physics and astrology in this respect. Still, correct predictions are not a possible criterion to distinguish sciences from astrology. While Newton's laws yield correct predictions, there are other academic fields that either do not make predictions at all or when they make predictions these are chronically wrong. Large parts of biology are not about predictions, but about descriptions of systems. A case in point is Monod's research on the lactose operon; there is no trace of prediction. It would be too harsh to dismiss Monod's work as non-science on a par with entrails-reading. Another example is economics. It makes predictions, but *quot caputa, tot sententiae*, as the Romans said, or there are as many predictions as there are economists. Even relatively short term projections of key-indicators of economic performance, such as the inflation rate or the change of GDP over a period of time are usually wrong. In fact, they are nearly always wrong. As far as the predictions are concerned, we may as well turn to entrails. Still, economics is a highly regarded social science. Economists would not identify themselves as soothsayers, but as proper researchers who develop and use rigorous methods to model economic performances. So, just because an activity makes correct predictions does not mean that it is a science, nor does the absence of correct predictions mean that it is not a science. We dismiss predictions as a possible criterion for science.

Another way to distinguish legitimate sciences from pseudo-sciences is their use of the scientific method. Reading entrails is not a scientific method, whereas making experiments with rolling balls on inclined surfaces is, hence Galilei was a scientist, but soothsayers are usually not. The trouble with this line of argument is that it does not really explain or define anything. We merely shift the problem from trying to say what science is to trying to describe what the scientific method is. This is probably not easier at all, but let us try anyway.

In order to understand the scientific method, we should start with the notion of the scientific theory. In colloquial usage the term "theory" is typically used to refer to intrinsically uncertain, speculative types of suppositions and ideas, often used with a distinct pejorative connotation, as in "xyz is pure theory." This is not the kind of theory that we mean in the context of science. Quite the opposite. In science theories are well established, profoundly researched and corroborated sets of ideas that collectively have some explanatory or predictive power. Einstein's theory of relativity is not speculative but it is a set of theoretical models intended to describe aspects of the world. Few people believe it is wrong, at least there is no evidence that it is. It may well turn out to be incorrect one day, but it certainly remains a very good approximation to any correct theory at least within the range of parameters for which it has been tested so far. Hence, theory is not a loose and informal, poorly thought through set of unfounded suppositions, but often the pinnacle of scientific achievement. The little brother of the scientific theory is the hypothesis. Hypotheses are conjectures about the state of the world, helper assumptions if you will. In the daily grind of scientific life they are frequently created and nearly as often rejected. Hypotheses come and go, but are very important vehicles to drive knowledge production in the daily routine of scientific practice. One can think of a theory as consisting of many well established hypotheses. Both hypotheses and theories do not exist in isolation, but they are constantly tested by experimental approaches. If a scientist comes up with a new hypothesis, another scientist will try to devise an experiment to test the hypothesis. This interplay between hypothesis and experiment is special for science and one possible way to define the scientific theory.

This process is very well illustrated by Monod's research. As he found out more about the mechanisms of the regulation of β-galactosidase, he started to formulate conjectures—hypotheses—about the system. For example, at some stage he posited that the regulation of genes takes place at the level of the gene. He did not know whether or not this was indeed the case, but the conjecture allowed Monod to design some experiments that tested it. Independently of the outcome of the experiments, whether they corroborated the hypothesis or rejected it, he could learn something from it. A hypothesis is valuable if it can be used to design insightful experiments. In the particular case of gene regulation, the gene removal experiment was such an experiment. It could not have been designed without having a specific hypothesis about the mechanisms of gene regulation. As it turned out, Monod's hypothesis was correct. More importantly, it was a useful question asked at the right time.

One person who tried to define the "scientific theory" was Anglo-Austrian philosopher of science Sir Karl Popper. He declared that hypotheses generation and testing is what makes science distinct from any other form of human activity. It is worthwhile to dwell a bit on his ideas because they are still widely held amongst scientists. According to Popper, testing hypotheses is very important, but not every type of test is equally powerful or informative. He pointed out that we know nothing about the truth or falsehood of hypotheses as long as our experiments are in agreement with them. No matter how often Newton's laws successfully pass an experimental verification, they may still be wrong. We only really learn something about nature when experiments fail to confirm a hypothesis. If the results of an experiment are not in agreement with the predictions of the hypothesis then we know that the hypothesis is incorrect. According to Popper at least, we can be certain that it is wrong. In a sense, when Monod conjectured that protein expression is regulated at the level of the gene, and his experiments did not contradict this assumption, he did not learn anything, at least nothing new.

Within this Popperian model of science, the scientific method is the permanent cycle of hypothesis creation and testing. No single hypothesis or theory is ever accepted as correct, but it is at best provisionally accepted, or provisionally believed in. Every such theory has the Damocles sword of eventual falsification hanging over it. In a way, it turns around Lewin's statement that we have to believe in Newton's laws. Popper wants scientists to disbelieve their very best theories. There can be no proof of these theories except the proof that they are wrong. This disbelief and systematic doubt is also what makes the difference between Newton and Vishnu, at least in Popper's world view. Just like religious beliefs, scientific theories can never be shown to be correct, but unlike religious beliefs they can be shown to be false. Popper defines a scientific theory or hypothesis as a system of thought from which experimentally testable predictions can be derived. If it is meant to be scientific, then it must allow us to make some sort of prediction that can be falsified. Scientific theories must actively seek to be tested, invite criticism. Each scientific theory has the next experiment as the provisional expiry date. Every test that has been passed is only a stay of execution, but never an acquittal.

Belief in Vishnu is not of this sort. There are no criteria that would allow any believer to devise an experiment at the end of which we could conclude that Vishnu does not exist. There is not a single religion that is falsifiable in this Popperian sense. For example, Luke 4:9-12 reports of the devil suggesting to Jesus to corroborate some predictions made in the scriptures:

And he [the devil] led Him [Jesus] to Jerusalem and had Him stand on the pinnacle of the temple, and said to Him, "If You are the Son of God, throw Yourself down from here; for it is written, 'he will command his angels concerning you to guard you,' and, 'on their hands they will bear you up, so that you will not strike your foot against a stone.' " And Jesus answered and said to him, "It is said, 'you shall not *put the Lord your God to the test*' "[Luke 4:9-12; emphasis added]

The devil was suggesting a so-called crucial experiment. Had Jesus indeed jumped and had he been saved by the angels, then this would have corroborated the scriptures, but it would by no means have proven them. On the other hand, had he not been saved, then this would have falsified the hypothesis that the scriptures are true. Jesus was of course entirely right in refusing to participate in an experiment that posed a significant risk of death for the participant (especially Jesus) and would not be approved by any credible ethics commission in the world.

The devil, however, may argue that the whole idea of the scripture is flawed. It pretends, he could say, to make predictions about events, promises about what happens when this or that is the case (for example save Jesus should he fall), but all this with an inbuilt proviso that it may not be put to a test. The system is self-immunising against a proper experimental evaluation which can be frustrating if one is of an empirical disposition. According to Popper this is not only frustrating, but also a clear sign that the scriptures are not reporting a scientific theory or hypothesis, but something else. Similar arguments can be made to argue that the Hindu Puranas and the belief in Vishnu are different from Newton's Principia Mathematica. By the same token, astrologers and soothsayers will not normally put themselves to falsification tests, at least they will not as a normal practice seek to be falsified. Whenever one of their predictions goes wrong then they will either have left town and cannot be held accountable for it, they may blame it on the entrails, or on some bad constellations of the stars. If the witchdoctor fails in his healing power, then this is because the ancestors did not approve of the patient, or the patient did not submit herself to the ancestors. Any failure to produce the correct outcome will be blamed on external factors ("the patient has a bad karma") and sometimes even used as a confirmation of the belief system, rather than a falsification.

That should settle the issue. The authority of science does not derive from it being certain, or proved beyond all doubt. Indeed science can be wrong, and held beliefs in science often turn out to be wrong. Science is different because its statements are falsifiable, and scientists consistently seek to falsify and be falsified, quite unlike religion or indeed any other human activity. Most working scientists, to the extent that they think about these issues at all, would say that falsifiability is a hallmark of science, not only good science, but any science.

The Popperian view of science has been hugely influential. To this day it defines for many scientists the essence of what makes their particular work a valid source of knowledge. It also comes with a number of assumptions, or rather conditions and limitations on what scientists can and cannot do. One requirement for a Popperian science to work is that falsifications (or corroborations) can be reproduced by anybody, at least in principle. The whole idea of a Popperian scientific method does not make any sense if, for example, only a special class of priests can test hypotheses, or only some anointed chosen ones can do this. The experiment must be public and accessible to everyone. This means that private knowledge, visions, revelations are all excluded as valid sources of falsifications. Any experiment by Prof. A must be reproducible for Prof. B, so both can agree or disagree on the conclusions that are drawn from these experiments.

Closely connected to this is the requirement that scientific theories and experimental tests are independent of the personal beliefs and values of the scientists who performed them. It should not matter what Newton's favourite dish was, what his religious beliefs were, how he felt about the politics of his day or whether he liked roses more than lilies. Personal tastes and values are not, should not be part of scientific thinking. So, falsifications must be reproducible by everyone and they must be independent of the values and beliefs that individual scientists hold. Falsification, and by extension science should be inter-subjective and inter-cultural. It should not matter where you come from. The science you produce is completely independent of your background and free from the values that the scientists themselves hold.[3]

Crossing the Styx

The ancient Greeks believed that all mortals, once they died, had to make the journey to the underworld, ruled by the god Hades. The border between the shadowy kingdom of Hades and the realm of the living was the river Styx. Once the grim ferryman Charon took the poor souls to the shore of the dead, there was no more escape back to life. The Styx marks the point of no return.

Imagine now that the Popperian account has convinced you and science is indeed a superior way of producing knowledge. Better and more reliable than personal revelations, more useful than oracles and more trustworthy than soothsayers and the whole lot. Let us assume for the time being that science gives objective facts that are probably not certain, but have been subject to a battery of rigorous tests, so that any unsubstantiated beliefs can be discovered and removed. You may then also value or admire the dispassionate approach of scientists who propose a hypothesis, make all the details available to the public so that any statement or claim can be subject to rigorous experimental tests and possible falsifications. Science at its purest is the pinnacle of human achievement.

This is how scientists like to think about their craft. They like to tell the tale of the dispassionate and disinterested quest for truth. But, is it all true? A closer look at science and how it interacts with us suggests that this may not be the case. It is probably, no certainly true that the outcome of an experiment is not dependent on the personal beliefs of the scientist, or whether he has a mistress or not. However, the choice of what experiment to perform and whether or not he believes that having a mistress is acceptable may depend on previous scientific outcomes. The point is that the moment we start to accept that scientific knowledge has a privileged status compared to other forms of knowledge, then this has tangible consequences for all, not only for scientists themselves. It will influence the choices we make in our life and what we expect, believe and accept as true. This has reverberations that go far beyond a mere philosophical commitment.

If someone falls ill then they normally seek the assistance of professional help. Assume the sufferer is given the choice between a proper university educated physician who uses the latest scientific knowledge, or a witchdoctor who uses ancient, yet unsubstantiated knowledge. Would she go for someone who bases his treatments on well tested approaches and falsifiable hypotheses or someone who does not. The particular choice will be largely dictated by personal beliefs about science and sorcery. In the western world most people have sufficient trust in science, even though this may be unreflected, so that they would choose the university medic, rather than the witchdoctor. Although many would probably also hedge their bets and use the services of non-scientific medicine, such as homeopathy, Chinese Traditional Medicine and so on.

In practice patients often do not have a free choice about their treatment options. Specifically in countries with socialised medicine such as the UK (which organises its public health service along Soviet style command economy principles) it is essentially the public purse that pays for the treatments of its citizens. Medicine is expensive and the treatments on offer are limited by a finite budget. Not everything can be paid for. Those who make the decision about what can and cannot be funded have to find some criteria to make their choices. Clearly, they cannot just make decisions based on personal preference, but they have to be accountable to the tax-payer who will scrutinise the way in which they reached their conclusions. Subjective likes and dislikes must not enter into this, but resources must be used so as to maximise overall value for money. This poses an immediate problem: How can one maximise value for money?

To figure this out the government in Britain has set up the National Institute for Clinical Excellence (NICE). Its responsibility is to review all available scientific evidence for efficaciousness of treatments. This assessment is mainly based on a search of the scientific medical literature. In this way NICE produces guidelines that detail for every possible condition, what types of treatments are available and which ones are recommended based on a benefit and cost calculation. It is by no means given that every efficacious treatment is recommended. There may be a drug or a treatment that prolongs the life of a sufferer, but when it is too expensive then NICE denies the treatment. The public will not fund it.

The details of how this is done are complicated, the principle is interesting though. The idea is that every treatment brings a certain benefit. This could be the reduction of pain, total cure from a condition or the prolonging of life. At the same time, any treatment also comes at a specific cost. Taking up a hospital bed, drugs, surgeries, all these things cost money. Given that there is only a limited amount of money available, the benefit and the cost of every medical intervention need to be weighed up against each other. In practice, this is a complicated process. It is not as simple as saying that a treatment will be funded if its costs are lower than the benefit. Any decision must also take into account overall demand, so it is more like a prioritisation process. There is a finite pot of money that can be spent, and when this is used up no treatment can be given any more. So, over the lifetime of a budget treatments must be made available based on their benefits and costs but also based on the total demand for healthcare and the size of the pot of money that is available.

The main difficulty of NICE's job is to come up with a reasonable conversion of the intangible and often subjective benefits gained from a treatment (which is usually experienced as an increase of life quality, less pain, cure from disease etc..) into a specific amount of pounds so as to be able to compare objectively benefit and cost. The value of reducing a patient's pain by a certain amount is difficult to quantify. It is not possible to convert money into pain, or pain relief. However, this conversion has to be done if the benefit of a treatment is to be quantified somehow. This leads to potentially awkward questions. How much is it worth to be relieved from the symptoms of the common cold, say. In a sense, we give partial answers to this question whenever we go to the chemist and buy pain-killers or cold medicines. A pack of Aspirin can be bought for as little as 40p. Most people are willing to pay this small amount of money to get some relief from head-ache or fever. Premium brand cold medicines can be significantly more expensive, costing around £5 in the supermarket. Still, there is a significant market for that, showing that relief from the symptoms of a cold is worth some money. However, if the price for these drugs would suddenly rise to £200 or even £1000, most would probably prefer to suffer a bit more from the cold than to spend such amounts of money. Naturally, the amount of money we are willing to spend on a treatment will depend on our circumstances and will also vary with the disease.

The people who professionally deal with these questions,

health economists, could base their quantification of the value of a treatment on the amount of money that an average person would be willing to pay to get its benefit. For the common cold this would be somewhere between 40p and £200. It is clear that this approach has a number of problems. Different people would certainly be willing to spend different amounts. The value will also depend on the wealth of the person and how significant the price of the drug is for the person in relation to her income. Then, in certain situations one would be willing to spend a bit more than in others. Imagine you suffer from a cold, but you have to perform well in an interview for a £1m job. You would certainly be inclined to spend much more on a drug than if the next day is just another routine day.

Let us consider a different, more fundamental problem illustrated by another scenario: Assume there is a treatment for condition X that can prolong the life of the sufferer, on average, by 2 years. How much is that worth? One million pounds or one hundred thousand or 10 million pounds? How about if the treatment prolongs life by 1 week, or a few days? How about if the treatment prolongs life, but at the cost of inflicting pain and an extremely poor quality of life for the patient? How can all these cases be quantified in an objective way? Most of us, who are not trained in health economics, are reluctant to address these questions. We reach the limits of quantification. It may make sense to ask somebody how much it is worth to her to have a certain amount of discomfort eased, but asking somebody how much her own life is worth in money is meaningless. The moment one dies the money itself loses all its value and utility. In this sense it is impossible to value one's own life in monetary terms.

For NICE such questions are unavoidable. They have to quantify. And they have to quantify based on scientific knowledge. Consider the alternatives. The NICE health economists could just toss coins. Whenever somebody needs treatment, the patient has to call heads or tails. A health economist tosses a 50p coin. If the right side comes up, treatment is granted. Otherwise, the patient has to pay herself or suffer in silence. Such a system is clearly unacceptable. It would not only be unfair, but also inefficient. Many people who are very ill but could be treated effectively will lose out during the coin tossing, but then many whose conditions have light symptoms only, but would be hard to treat would win. Overall, not the maximum amount of suffering would be alleviated.

Alternatively, the health economists could apply their own judgment to each case. They may think that disease X is worse than disease Y, so should always be prioritised. If they ranked every disease in this way, then they could come up with a list of priorities that could inform funding decisions. Clearly, this is not a good system either, because it entirely relies on opinions of individuals. One could extend this and have the public vote on this ranking. This would be democratic, but has its own problems. For one, most people could not be expected to understand most conditions that they have to vote for. Moreover, the sheer number of conditions to rank is very high. The voting process would exhaust most citizens who have other things to do. Direct democracy is not a good solution in this case.

It seems that quantifying benefit and comparing it to costs is the least worst thing to do. Once we agree on that, then a decision has to be made as to how suffering and illness can be quantified and most of all to what extent various treatments are efficacious and what they cost. How should this be done? Well, scientific studies can tell us. Objective knowledge based on publicly available data should scrutinise the effect of every possible treatment, so that NICE can get an overview over how much an intervention would cost and how well it would work. Whether or not this scientific knowledge is right or certain is not the point. The important thing is that scientific knowledge is again based on objective, reproducible tangibles that can be scrutinised by everyone.

It seems that once it is decided to use scientific knowledge to make decisions on treatments, then this invariably leads to a point where life and death decisions are based on a weighing up of current scientific information and available budgets. The health economists have no choice but to remain on the scientific path. Failing to limit the use of treatments in some cases would significantly distort the quality of the health service as a whole. So, the only choice for NICE is to specify how much a week of life under these or those conditions is worth in amounts of money. Consequently, the decision to grant or deny treatment to somebody, to extend somebody's life or to cut it short becomes an exercise in number-crunching, a cost-benefit analysis. Whenever they left the path of science, then they would always be vulnerable to objections and lose their ability to objectively justify their choices. Once the Styx is crossed, there is no way back.

It may well be that science as long as it remains within the confines of the laboratory of the curious researcher is indeed value free. Probably, Monod's bacteria and Galilei's inclined surfaces do not depend on values, at least not on moral and cultural values. However, things change once scientific thinking propagates out into the realm of the public. When science and its results start to interact with society, then the value-freeness is quickly gone. In the case of NICE, the scientific outlook forced a particular type of value choice on society, a new economics of suffering. The scientific approach that NICE takes to optimise the use of funds leads to a very particular way of addressing certain types of ethical questions. "How much is it worth to alleviate a certain amount of suffering given certain circumstances?" NICE has to find an answer to that and it has to find an answer in a way that is transparent and reproducible, just as every scientist has to report the findings of her experiments in a way that is reproducible. There is only one way for NICE to be objective and transparent: They need to rely on quality knowledge about diseases and their treatments. This means one thing only: They need to use scientific knowledge to inform their decisions.

The question is not so much whether or not this is a good or bad way to make decisions. It is the only way. At least the only acceptable way to allocate limited funds. Scientific advice cannot be brushed off without risking the charge of irrationality. We cannot decide *not* to use available scientific research to influence the organisation of public healthcare. It would be absurd to do so. Yet, the moment we do take into account research then this has a tangible consequence. Conditions and their cure must be assigned monetary values based on sound scientific evidence. Life and death decisions have to be reduced to numbers. Yet, every such decision is also a value choice. We decide whether or not it is worth to help a sufferer or not.

Clearly, if we absolutely wanted to, we could decide to organise health systems along very different principles, using different values. However, given the scientific insights we have, given the highly sophisticated research organisations that we have created, along with the entire infra-structure of knowledge dissemination and creation, and given that we believe in science, it seems absurd to do anything but base decisions on science. In this sense, we do not really have a choice. The scientific perspective is like crossing the Styx. Once one has taken it, once one has accepted that everything needs to be accounted for and justified then it becomes impossible to turn back. The moment one starts to pay attention to scientific evidence for treatment X or treatment Y it becomes completely absurd to ignore it. If we know that one treatment is better than another, then there is no reason to choose the worse over the better one. By the same token, once we know precisely what the effects of a treatment are and what its costs are, then health-economics follows nearly automatically and it comes with its own ethics. Charon the ferryman does not take us back.

One can argue that it may not even be desirable to be taken back. Why should we use inferior types of knowledge, non-scientific knowledge, when we do have properly tested scientific knowledge available? Would this not return us to the dark ages when our thinking was poisoned by superstition? A least the new prophets of scientific rationalism want us to believe that there is no knowledge but scientific knowledge. If it is not tried and tested and falsifiable, then it is not knowledge of any value. Only scientific knowledge is worth knowing.

Shamans and healers

There are many types of knowledge we have that are non-scientific, but clearly worth knowing. I know, for example, that I do not like fish and seafood. This knowledge is not scientifically corroborated, nor falsifiable, I just know it and behave accordingly. I also know how to drive a car, but this is again not based on scientific principles, it's a skill that I have acquired. Similarly, bushmen in the Namib dessert know how to find water in the most arid environments. Again, this is knowledge that is passed down from generation to generation. This does not mean that their methods cannot be evaluated scientifically. They clearly can. Yet, it shows that there are types of knowing that are non-rigorous that have been found by trial an error and they are still legitimate, indeed essential.

There is no doubt that scientific medicine has led to great advances. It can treat conditions that were untreatable before and reduces suffering where suffering could not have been reduced before. It would be unreasonable to argue against that. At the same time, when modern research-based medicine derives its power from science, science itself can also limit it. A deep methodological problem of medical research is that the subject of medical research, the individual patient, is not an idealised system and difficult to do research on. Even when compared to, say, microbiology, where the research subjects are bacteria or yeast cells, it is much more difficult to obtain reliable data about humans. People are complex and know that they are being researched upon and may change their behaviour in ways that may affect the outcome of the research. One aspect of this changed behaviour is the well-known placebo effect: Patients who receive treatment, say a drug, often show an effect even if the drug has no active ingredient at all. The mere fact that they receive treatment and believe the treatment to be effective is sometimes sufficient. The effects of placebos are by no means imagined, but can can be concrete physiological changes that are themselves objectively verifiable.

In medical and pharmaceutical research the placebo effect is mostly considered a problem particularly in the context of testing the efficaciousness of drugs. Naively one could think that testing a drug is a simple matter of giving a putatively efficient compound to a group of patients and then checking whether or not they get better. In reality the placebo effect means that patients may well get better just because they have an expectation that this will happen. The effect is significant and must be somehow factored out in order to understand the "true" efficacy of the drug. In practice this is done by using double blind randomised trials. In these the patients are divided up randomly into two groups. One group receives only sugar-pills or any other inert substance and the other group receives the real compound to be tested. The patients themselves do not know in which group they are, neither do the physicians that are in direct contact with the patients. Usually in such trials one can observe an improvement of the condition in both cases. The effect seen in the placebo group is usually discounted and taken as a basic baseline of "no effect." The efficaciousness of the active ingredient or indeed any type of treatment is then measured by the amount of improvement seen relative to the placebo group. A treatment or a drug are only scientifically shown to be efficacious if they pass this test, that is if they are demonstrably doing better than a placebo comparison. Otherwise, the drug or treatment are considered non-efficacious.

Double blind randomised trials of this sort have become the gold standard in medical sciences. There is an undeniable logic to them. It is certainly not worth paying money for compounds or treatments that apparently do nothing. Particularly in countries with socialised medicine it makes sense that the public purse does not fund treatments that are not efficacious. When resources are limited it is imperative to squeeze out the maximum of every penny being spent. It is therefore not surprising that publicly funded health-systems come up with organisations such as NICE that assess every treatment according to scientific principles.

It is not only the publicly funded health service that has an internal drive towards scientific approaches, but nearly every stakeholder in the system has an interest to follow proper evidence-based medicine. Physicians themselves certainly have a strong interest in being seen as scientific. If nothing else, with their years of university training it would be difficult to accept that shamans and witchdoctors could do equally well as they do. Would a doctor not be reduced to a mere healer if the cure of the patient was not due to her skills but due to some spooky placebo effect? The same is true for the pharmaceutical companies. What if it came out that the drugs that cost millions of dollars to develop are not much better than a simple sugar pill? And for the rest of us, why should we pay for a treatment that is based on our own powers alone? Would it not amount to quackery to sell placebos? We should take the placebo effect gratefully, and then insist on a a drug or treatment that can be shown to do better than that. It would be irrational to be content with the second best, the less efficient, not to insist on drugs and treatments that give effects over and above the placebo.

As much as accepting science-based medicine seems self-evidently the right thing to do, at a second glance things are a bit more complex than first suspected. One of the underlying assumptions about the treatment process is that a patient has a diagnosed condition X and the physician prescribes one of several available treatments for X. The idea is that among those treatment options some are more effective than others, as determined by scientific studies. Normally, the physician chooses the one treatment that promises the best outcome based on available scientific evidence. Real scenarios may be very different. Not all diseases are specific, many remain undiagnosed and treatment is often ineffective. This is especially true for relatively minor conditions such as chronic digestive disorders, headaches, arthritis, asthma, light depressions and the like. These ailments can drag on for years without any satisfactory solution. They are not life-threatening diseases, but often incurable in the sense that there is no effective treatment for them. Many of those disorders may not even correspond to a defined condition that could be identified in a larger class of patients and may even be unique to the patient. For the sufferer they can still imply a significant reduction in life quality and even impact on the patient's economic activity. So, altogether they are significant.

When the condition is not clearly identifiable, then it is a moving target for scientific research. If the condition is personal to the sufferer, then certainly no research is available (because nobody else has exactly the same problem). Scientific evidence cannot be produced because the single patient cannot be divided into two controlled groups. The underlying mechanisms that cause the disease or disorder are nearly certainly unknown. In those cases the whole idea of evidence-based medicine becomes irrelevant because there is no evidence to cover the particular case. Science encounters a methodological limit. Medicine is powerless. We can fly to the moon (although it seems that we have actually forgotten how to do this), but we cannot cure a chronic headache. The only option is to fight the symptoms.

In such cases, it is at least conceivable that non-scientific approaches to medicine can help. They certainly cannot be any worse because science-based medicine does not do anything. Many non-specific ailments may well have social or psychological root causes. It is not at all implausible that a healer or a shaman can address those more effectively than evidence-based medicine. This is certainly true for culture specific conditions, such as for example "being obsessed by an evil spirit." This seems laughable from a western perspective, but is a real and debilitating condition in some communities. There is no NICE guideline for this, but the witchdoctor in the village has a way of dealing with that. Another condition that seems to be culture specific to Germans is the so-called "Kreislaufschwäche" which literally translates to "weakness of the circulation." Go to any German chemist and they have lots of little medicines for this condition. In the English speaking world your pharmacist would probably not know what you mean.

More mundane conditions such as chronic headaches are probably not so very different. It could well be that alternative medicines can help in these cases. The problem is that they are difficult to evaluate scientifically because these alternative approaches often rely on a close patient-healer interaction, whereas in scientific medicine the effect of the treatment is usually considered independent of the particular social settings in which the treatments are administered. There are therefore fundamental methodological problems to assess things like homeopathy, witchcraft or chiropractice because their procedures may not be amenable for double blind randomised trials. It is difficult to find the sugar-pill equivalent of things like shaman ceremonies. In practice one would have to devise a sham-shaman-ceremony, that is a pseudo-ritual that to the patient looks and feels the same as the real ceremony, but changes the correct ritual in crucial ways so as to render it ineffective. Shaman rituals often involve a ritual where the healer goes into a state of trance that allows him to visit the land of the ancestors. They use animal bones or charms in their treatments along with herbal medicines.

So in a sham ritual the shaman could use the wrong type of bone or a "random" charm. He could also perform his dances in the wrong order or just pretend to visit the ancestors' land. Any variation of this sort would be fine for the sham treatment, as long as it breaks some essential rules of good shamanism.

Let us assume that a proper study of shaman treatments with control groups and all the bells and whistles of good science has been performed. There are then two possible outcomes. Either, the sham treatment turns out to be worse, that is less effective, than the proper shaman treatment. Alternatively, it could turn out that there is no difference. In the latter case, the verdict is clear, at least from a scientific point of view. Shamanism is not better than a placebo for treating condition X and has to be rejected. In the first case one may be tempted to conclude that shamanism if done properly is actually better than a placebo. Unfortunately, on second thoughts this is not so clear. The first target for criticism would be the experimental setup of the study itself. Certainly the shaman himself would have known whether he performed a sham ceremony or a proper one. It is well known from drug trials that the expectation and the knowledge of the therapist play an important part in the outcome of the treatment. This is also the reason why in drug trials the people who are in contact with the patients themselves do not know whether a particular patient is in the placebo group or whether she gets the proper drug. Based on this we would have to conclude that the difference between shaman treatment and sham-shaman treatment, if there was one indeed, is also placebo. Either way, the evidence is inconclusive. Similar kinds of arguments can be made for chiropractors, homeopaths, witch-healers and the whole lot. The criteria used in scientific medicine cannot be applied to the village witch-doctor. For all parties, including patients, it therefore becomes very difficult to accept these alternative treatments. There is no way back from the Styx.

This hypothetical test indicates a much deeper problem when

applying scientific principles to real life situations such as medical treatments. In order to be able to draw any conclusions from an experiment it is essential to control the environment and to exclude environmental variation and noise from the experimental system. This is why Galilei built his inclined planes and Monod went to great lengths to construct variants of *E.coli* that allowed him to address one specific variable at a time. Double blind randomised trials are meant to be the medical equivalents of inclined planes and double mutants of *E.coli*. They are explicitly designed so as to allow testing the effect of one and only one variable, namely the active ingredient. Moreover, they are designed to test the physiological effect of the drug and to separate out any second order effects that come from the physician-patient relationship, the setting of the test and the expectations of the patients. In that sense, these trials attempt to realise idealised conditions.

There are a number of assumptions built into scientific medicine and its evaluation and testing. One assumption built into trials is that the physical and bio-chemical effect of the treatment is separable from the therapist-patient relationship (which is thought of as merely psychological). This assumption is certainly not true in the case of shamans, and many other forms of non-scientific medicine, where inter-person relationships are usually considered crucial. Another assumption is that the placebo-effect is a non-effect.

In practice, these assumptions have tangible consequences. Certain types of treatments are permanently excluded from the list of acceptable procedures, because they cannot be tested using double blind tests or because they rely on the the placebo effect. It is built into the practice of medical research that the evidence for shamanism cannot be better than inconclusive. This does not, however, mean that it is not doing any good. For those who are not familiar with shaman practice the rituals they perform seem rather far-fetched. Most of the people grown up in a western tradition will struggle to find any sympathy for it. Still, in many countries, particularly in Africa, shamanism (or sangomas as they are called) are considered a true alternative to science-based medicine in some cases (Cook, 2009). In certain contexts shamans fulfil important roles and their treatment works, at least sometimes.

Using the criteria of medical science to systematically test shaman treatments risks missing the point. The important thing of shaman medicine is precisely the cultural context and the ritual associated with it. The "active ingredient" so to speak are not the venerated ancestors, but the shaman-patient relationship and the patient's belief in the ritual. In a sense, together shaman and patient create powerful placebos. Any scientific assessment of shaman treatments would necessarily try to identify active ingredients. Since there are none, they are bound to fail. From a scientific perspective one is therefore led to conclude that this is only a zero-effect treatment, unless one can show a direct link between ancestors and physical changes in the body, which will not be possible. Still, people go to shamans, so they must get some benefit from it. Scientifically assessing shamanism and similar endeavours is an attempt at trans-Styxian communication, which is difficult to get right.

Similar arguments are also applicable to other forms of non-scientific medicine, such as homeopathy, chiropractice and so on. In all these fields the relationship of the "healer" with the patient is a crucial factor for the success of the treatment. What precisely the healer is doing is perhaps not as important as the fact that whatever he or she is doing makes cultural sense and fits with the expectations of the patient. Now, if the ritual is important for non-scientific medicine, then there is no reason to assume that it does not have its place in scientific medicine as well. There will be no calling of the ancestors in modern hospitals, nor will a heart surgeon cast the chicken-bone, but there is still ritual involved. Seeing a doctor has a specific pattern to it that one could consider as a ritual. There is the waiting room that gets the patient into the mood. Health information posters on the wall and leaflets give subliminal information. Then the patient is called into the treatment room which invariably has a bench, a privacy curtain and a display of medical tools such as syringes, tubes, stethoscopes. In many countries there is the white coat of the practitioner and maybe the degree certificate on the wall. We do not normally think of these elements as a ritual quite in the sense that we would see the shaman dance. Still, it has the same effect (Kaptchuk, 2008; Miller and Kaptchuk, 2002).

Fabrizio Benedetto and his group at the University of Turin in Italy have done some systematic research into the importance of ritual in western medicine (Benedetto et al., 2004). To do this they devised a very interesting variation on the normal double blind experiments common in medicine. Instead of measuring the effectiveness of a drug, as is usual in such experiments, they set out to measure the importance of the ritual surrounding the administration of the drug. In their experiments they split up their patients into two groups. In standard double blind experiments only one group would receive a drug, whereas the other group receives a placebo. Not so in Benedetti's experiments where *both* groups received an active compound. The difference is that only one group (the "overt" group) was told so, whereas the "hidden" group did not know that they received a drug. Apart from that the treatment between the two groups did not differ. In one of their experiments Benedetti's group tested the effect of common painkillers in patients right after surgery. Patients in the overt group were told by a doctor that they received a powerful drug, whereas the hidden group got the injection without being told so and without a doctor or nurse being in the room (the patients were all on a drip that could be programmed to administer a certain dosage at a given time). Following the injection of the drug, the patients were asked to rate their pain levels. What Benedetti found was that the pain in the hidden group who did not know that they got a drug subsided slower than in the overt group and the long-term pain reduction was altogether smaller, although only marginally so.

More interesting was the second experiment they performed. Now, instead of determining pain levels in groups that knew/did not know that they got a pain killer, they told both groups that they administered a drug. However, after some time they stopped injecting the compound, but only informed patients in the overt group. The hidden group was not informed that they no longer got the drug. Once they ceased delivering the drug there was an immediate and sustained increase of the pain levels in the overt group. In the hidden group, whose treatment was likewise interrupted the pain-levels remained unchanged and much lower than in the overt group.

Benedetti observed a much stronger effect when he tried the same with diazepam (valium) a strong sedative and anti-depressant. When he used this to treat patients with anxiety, the overt group showed a very strong response to the drug, whereas in the hidden group the drug had no effect. Similarly, when they hid the interruption of the treatment from the patients, then they had no significant relapse in anxiety, whereas there was a strong relapse when the patients knew that they no longer received any drugs.

These experiments indicate that even in the context of a modern western hospital the outcomes of a treatment cannot be merely reduced to the narrowly defined effects of a drug, or the specific physical effect of a treatment. Benedetti's experiments only just started to investigate the power of ritual in western medicine. At the very least he shows that there probably is much more to be investigated. It may not be that far-fetched to compare modern health services with witch doctors and shamans. At least some of the effects of treatments are due to the ritual, or the authority of the treating agent. In some cases (diazepam) the drug is a mere symbol of healing with apparently no specific effect if it is detached from the treatment ritual. A drug can work because the doctor tells you so, and it can stop working because the doctor tells you so. If you do not know you get it, it may not be as powerful, or even be completely ineffective.

The Benedetti experiments probably only show the tip of the iceberg. Who knows what hidden relationships between carers and patients, their expectations and the ritual they are performing influences the effectiveness of treatments. What is so far known from research into the placebo effect suggests that there could be much more than one thinks. However, scientifically these subtleties are difficult to study. Understanding ritual in western medicine may well hit the same methodological barriers that limit research into shamanism.

Additionally, there is another problem: Reflexivity. Imagine now for the moment that a group found that treatment Y only works in conjunction with a ritual, or that interaction pattern Z makes patience better even if they do not receive any effective drugs. If the results of this research become widely known and patients start to recognise the pattern of interaction, then it may well be that the patients lose belief in the efficacy of medical intervention. Even though most patients have only a vague understanding of scientific principles, their belief in science is probably still an important part of the healing process. Once this is broken, who knows how effective treatments remain. For this reason, it is not even desirable to find out too much about the power of ritual in western medicine, because it may undermine the very power it studies.

How to falsify a climate model

Medical research is just one example of science that propagates out from its immediate domain and raises moral issues. The same applies to many other types of research. Science is not merely a structured way of making arguments, but it penetrates into all crevices of our life, changing how we think about things, even how we are supposed to feel about things. We allow this to happen because there is a general sense that science produces reliable knowledge, or at least knowledge that is much better than knowledge that is produced by other types of human activities. The question we asked at the beginning of this chapter was how we can justify this belief rationally. Why does science have this apparently privileged position compared to other forms of knowledge? Why do we let it rule our lives to the extent we do? The most promising explanation we found so far was Popper's criterion of falsification, that is the idea that scientific statements, and only those, can be falsified by experiments.

Popper's philosophy and his criterion has been very influential, particularly among working scientists. This is somewhat surprising, because a closer look reveals that research rarely fits well into the Popperian scheme of things. Take medical research as an example. It is not straightforward to understand in what sense double-blind trials falsify theories. Popper's philosophy of science makes most sense in the context of speculative theories that propose specific claims about the nature of reality. Einstein's theory of relativity is a case in point, as is Newton's physics. Double blind drug tests in medical research are somewhat different, because they are not designed to test theories. The hypothesis that drug X is effective is an empirical question, but it is not about the nature of reality. It is more like the question whether the grandmother is in the room next door or not. This latter question can be easily verified: Go and check, then you know. Similarly with drugs. If you want to know whether an active ingredient is efficacious, just try it out. With a bit of imagination it is perhaps possible to shoehorn medical research into Popper's schemes, but there will be no natural fit. There are, however, other types of sciences where even shoehorning does not work so well any longer. Interestingly, or worryingly, these sciences are often those that are of highest practical importance.

One example is climate change science. The world-climate is heating up, which is most likely caused by human behaviours. If the climate gets warmer a bit, it probably does not matter, but if it heats up a lot, then this could have terrible consequences. At the very least it would cost a lot of money. Something needs to be done. The question is what and by whom. Decisions must be made. The dilemma is that taking actions against climate change also costs a lot of money in the short term. It requires significant political capital to be expended by the leaders of the day. Therefore, there are two questions that need to be addressed. Is it indeed the case that climate change is predominantly or at least significantly anthropogenic, that is caused by human activities? If so, what is the optimal strategy of avoiding unacceptable or catastrophic long-term outcomes? Ultimately, this comes down to understanding how much climate measures cost now, what their precise consequences would be and how much they would save in the long-run compared to no action.

Politicians, being experts in the affairs of the state and not in climate science are clearly not in a position to understand the consequences of their actions. They need to ask experts for advice, be it soothsayers, clairvoyants or scientists. The choice between the three is easy to make in this case. What is clearly needed are well founded forecasts and the ability to understand what the consequences of various scenarios and options for actions are. In other words: We need models. Making models of the climate is no easy task. There is a whole host of things that influence the climate. All these effects must be somehow specified. Then computers can be used to process all the available information and create concrete predictions.

Any attempt at modelling must start with an understanding of the processes that generate the climate in the absence of human interventions. The dominating effect of the climate is the amount of radiation originating from the sun, the primary source of energy that powers the entire climate system of the planet. The intensity of this radiation is not equal across all parts of the planet and varies geographically and with the seasons. The climatic patterns in different parts of the world are caused by these variations in sun-energy and how it is shifted around internally and eventually emitted back into space again.

One way in which sun-energy, or heat differences between regions on the planet are equilibrated are currents of air. Roughly speaking air masses are shifted from the equatorial regions to the poles. The oceans also re-distribute heat. As water is heated up it starts to flow around the planet forming the great weather patterns. The Benguela current in southern Africa is an example, another one is the Gulf stream that keeps European winters tolerable by adding some Mexican heat. These and other currents are driven by differences in salinity and temperature, wind or water density differentials.

These effects are just the bare-bone ingredients of a proper climate model. Much more is needed for a proper model. In order to be able to represent the movement of air currents a model must represent the equator and the poles and everything in-between. It must be able to account for air-flows between the regions in response to solar radiation. So, ideally, the model represents every point on the planet, but this is not practically possible. In practice it is necessary to "chop up" the planet into small regions, say small squares of a few miles which can then be regarded as homogenous. Such a model would probably not distinguish between Paris and London. Both would be, for climate purposes, the same place, always receive the same amount of sun and have the same weather at any time. We know that this is not true, but for the purpose of the climate model this approximation to reality may well be good enough. The larger the regions, the easier it is to run the climate model and to extract some information from it. In this respect, larger squares are better, but there is a downside to them. A coarse chopping makes the model less accurate, probably even dramatically so when the squares get too large. Mountain ranges, for example, change the way in which air can flow. In certain areas they may be an effective barrier for clouds and there is a very different weather cis and trans this range. Large bodies of water are important, as are large islands in that they change the way in which heat is absorbed and radiated back. Coarse graining will wash out these topological features, and makes the model wrong. Perhaps the resulting error is acceptable when Paris and London are considered the same place, but it would certainly go too far if Paris and Bamako were lumped into the same square. That would not work.

Coarse graining has to be done for every aspect of the model. The atmosphere is not homogenous, and there are air movements— vertical as well as horizontal. So, modellers chop up the atmosphere into various discrete vertical layers. Mountains and forests and big bodies of water change the way heat is absorbed. Again, not every tree can be represented, so the model must coarse grain into different types of landscapes. All this simplifying and coarse graining requires great skill to get right. The modeller needs to strike the right balance between simply capturing all the important features while keeping the model feasible. Compromises must be made and corners need to be cut to get there. The key to success in this is to find the right simplifications, to omit enough to make the model feasible, but to keep just enough so that the predictions that come out from the model are correct. To make the right choice is the skill of the modeller.

In reality it is even harder to do this than the description suggests so far. In climate models elements within the model act back on either themselves and/or other parts of the model. This can be illustrated with clouds. The warmer the planet, the more water will evaporate from the oceans, which leads to the formation of clouds. Once these clouds are formed they tend to shield the surface of the planet from further solar radiation, reducing the temperature and also the rate of cloud formation. In this sense clouds act back on their own formation. This is not the whole story yet. Water vapour is also a climate gas, just like CO_2 and Methane. More water vapour in the atmosphere means higher average temperatures which means that more water vapour is created.

So, already in this simple example of water vapour there are two different feedback loops at work simultaneously. For the climate modeller it is very important to understand these effects because they materially change the predictions of the model. However, just knowing that such a feedback effect is present is not enough, it must also be known exactly how strong the effect is. In practice this boils down to being able to write down an equation that precisely specifies by how much the greenhouse effect is increased when the amount of of water vapour is increased by a given amount. Measuring these quantities is challenging. For some feedback loops it can be done, for others it cannot. The problem with feedback is that the system-behaviour can be very sensitive to the precise nature of the feed-back. Even a relatively minor mistake in specifying this can lead to dramatically wrong model outcomes. When a particular instance of feedback is sensitive and when it is not depends to a large extent on the context in which it occurs within the model. Luckily there are methods for modellers to find out whether or not this is the case. Unknown feedbacks, that is effects that are acting in the real atmosphere, but are not represented in the model, are more disconcerting. One can never be sure that a model takes into account all the relevant effects. And even with the effects that are included, one can never be sure that they are included correctly.

Climate models need to be coarse grained descriptions of reality. Not everything that is known can be included in the model, and not everything that should be included in the model is known. At the same time, when features that should be in are not, but features that should not be are in, or both, then this can have major effects. This is why climate modelling is so hard.

At a first glance the coarse graining that climate modellers have to do seems similar to what Galilei did. Removing complexity from the real phenomenon was the essential part of his strategy that ultimately led to modelling success. Galilei could only predict the fall of bodies because he made the crucial idealisation of ignoring friction. The simplifications of climate scientists are therefore seemingly in a good scientific tradition of the Galileian paradigm.

A closer examination shows that the case is more intricate than it first appears. It certainly is the case that simplification is a practical necessity both for Galilei and for climate science. Beyond that, it is not clear that there are any more similarities. What made Galilei's idealisation so powerful is that it revealed an essential core-effect that was concealed by noise. Superficially it appears as though the weight or shape of an object are relevant for the speed of the fall. In reality they are not, but their apparent importance is a result of disturbing effects. There is a core phenomenon *gravity* that can be modelled and thought about in isolation whence it will yield simple laws.

The climate is not like this at all. The hundreds of complicating factors that make the climate are the system itself. Strip them away and you lose the essence of the climate. There is no core to the system that reflects the simple fundamental nature of what the climate is. There is no idealised climate that is both simple and relevant as a model for the real climate. This does not mean that it is not possible to simplify and to leave out some effects. What it does mean is that there are many possible choices of what should and should not to be included. Some of these models will be good, in the sense that they make good predictions, some of the models will be not so good, in the sense that their predictions will be affected by a large error, but there is no single model that will shine as the best model. What the modeller chooses to include in her model will depend on the particular question at hand, the available computational resources, it will depend on her judgement as to what is important and what is not important, her knowledge and the time she is able to invest in making the model.

Choosing the right ingredients for the model is a problem, but it is not the only problem. There is another issue well known to all computer scientists, in fact everybody who has ever written computer code and unfortunately also most users of computers: Bugs. Apart from being models, climate models are most of all computer programs, indeed very large computer programs. Some of the people who programmed the models are very good at coding, some of them are probably not very good at it. In any case, even the best of all programmers make mistakes. Writing computer code means giving very precise sets of instructions to a machine. The machine then executes these commands ruthlessly, without applying any common sense. For the human programmer this means that she must specify in meticulous detail, with the highest precision what she wants the machine to do.

Thinking like this is difficult for us humans. In our daily dealings we are imprecise. When we give instructions most of the meaning of our instructions is implied. Understanding commands requires some common sense. Arguably, it is essential to be imprecise in normal daily contexts. Covering any possibility, and specifying in great detail what we mean would make communications not only lengthy, but most of all difficult to comprehend. An example can illustrate just how much is left out. In one of the lecture halls at the department of biosciences at my university there is a sign saying:

All phones must be switched off during lectures.

For the students (and indeed the lecturer) it is quite clear what this instruction means. Ringing phones disrupt the lectures, so students must make sure that their mobile phones are off. It is easy to understand the intention behind this sign and what it attempts to convey. However, this obvious meaning of the message of the sign is not at all communicated by the sign itself. If we forget about common sense and take the instruction literally, then it would mean something different: "**All** phones must be switched off during lectures" means that whenever a lecture is going on then all phones there are must be switched off. The scope of the devices to be turned off is not limited to the lecture hall and the quantifier "all" means all. Clearly, what the sign actually says and what it is intended to say are two very different things (it would be presumptuous of the University of Kent to order everybody to turn off their phone, just because there are lectures going on). Nearly all signs are imprecise like this and this is not a problem. The person who made the sign can assume that the readers are capable of interpreting signs in a reasonable way.

When it comes to programming computers, then it is no longer possible to assume such a common sense interpretation. If one had the equivalent of this kind of imprecision in a computer program, the program would indeed turn off all the mobile phones there are. It would not consult common sense and restrict the application of the rule to some implicitly specified set of instances, because it does not have any common sense. The result is a programming error. Since we (including good programmers) are not good at thinking in precise terms, programs often do things that were not intended by the programmer. This is not only very common in software but unavoidable. "Bugs," as they are called in the jargon, are a big problem in software engineering. Despite extensive tests and efforts to the contrary, there is not a single piece of new software that is released without an extensive set of bugs. Windows users of a bygone generation will know the "Blue Screen of Death," which always appeared when a bug was encountered in Windows. In other operating systems the same problem exists. Over time some of these error are eradicated, but some always remain.

Climate models are computer programs just as operating systems are. If anything, they are programmed to lower standards, often by scientists whose specialisation is not computer programming, but something else. They are very complex and so inevitably have bugs too. One would think that these bugs would sooner or later manifest themselves by a blue screen of death or a crash of the program. Indeed, some of them do, the easy bugs. Then there is a bug infestation that does not manifest itself in obvious ways, but remains undiscovered. The output looks ok, although there may be the crucial error hidden somewhere in a sub-routine half-way through a program with several hundreds of thousands of lines of code. Often these bugs are minor errors, where an assignment was not made correctly, an instruction that should be deleted has not been deleted, or a minus sign is in the wrong place. Innocuous as this may seem it can have a big impact on the result. Still, nobody may notice it or find it. Even when all the assumptions of the model are correct, then such bugs may still make the model wrong.

Bugs in the widest sense of the word are a big problem in any science that makes heavy use of computational models. There are strategies to validate a model and check that it is indeed performing as it should be, but these tests will rarely be sufficient to test the general reliability of the software. Particularly for more complicated models there is always a residual uncertainty about the correctness of the program.

There is a dilemma here for the modeller. On the one hand, there is no well defined Galileian essence that allows models of the climate that are both simple and relevant. On the other hand total models, that is models that include everything, are not feasible because we do not know all the effects. Even if we knew everything about the climate, bugs mean that the resulting model would be unreliable. Therefore, the modeller has to make choices as to what she wants to take into account and what she does not want to take into account. She must simplify. The resulting models will still be large, take a long time to run and ignore a number of known effects, but hopefully they will be good enough to be of some use. Therefore, climate models are a sea of uncertainty. Climate scientists are very well aware of this and would never put all their trust into a single model. Instead, they consider populations of models. They have a large number of independent programs that calculate the future climate, each of them slightly different.

One of the leading research groups in climate science is based at Oxford University in the UK. It alone lists 8 different models on its website[4]. Each of those models has a slightly different purpose, a different parameterisation and exists in many different versions, and each of the models produces somewhat different predictions. The question is this: If all these models are different, how can we know which one of those is the correct one? How can we be certain that any one of those is correct?

At least as long as we believe Popper and his definition of science then there is a very clear answer to this: We cannot be certain that any of these models is correct, or makes the correct prediction. We can only assume that the models are correct until there is some experiment to falsify its correctness. The mantra is that science is falsifiable, and when it is not falsifiable then it is not science.

The trouble is that this idea of falsification does not make any sense in this context. First of all, climate models are models and not scientific theories. They use scientific theories to come up with predictions, but they are not theories themselves. They are computational tools that allow scientists to synthesise all their insights and understanding, process it and draw conclusions from it, all computer aided. In this sense, the models are devices to process information or knowledge that is derived from scientific theories. If a climate model is falsified, then this is certainly not a falsification of all the theories and hypotheses that go into the model. It may not even be a falsification of one particular theory or hypothesis. All the theoretical ingredients that went into the model may well have been correct, but they have probably been combined in the wrong way. Maybe one of the feedback loops was not described correctly, or there is an effect that has been omitted when it was in fact important, or maybe the coarse graining of the model was too coarse and something else was inaccurate. If a climate model proves to be wrong, then this does not falsify any theory. It only means the model was wrong. Nothing more, nothing less.

In this sense, falsification does not apply to climate science. Beyond this rather theoretical point there is a much more urgent, and practically relevant issue. Not only is it true that falsification of climate models would not falsify any theory in Popper's sense, but the models themselves are untestable. This *is* worrying in practice and an Achilles heel of climate science. Say a particular climate model predicts that the temperature increases by so and so many degrees under a specific set of circumstances (so and so much carbon has been released per year) by 2050. How can we possibly check that the model is correct? The only way to do this is to wait until 2050 and measure the temperature then. The obvious frustration about this procedure is the long time lag. Most of the climate scientists active today will probably not want to wait for so long. We conclude: Climate models are not falsifiable, unless you have a lot of time on your hands.

The real problem, however, is that even if one wanted to wait so long it would completely miss the point of the models themselves. The reason is as follows: Climate models are developed to provide scenarios of the future in order to find out whether these should be avoided or desired. We need to know the effect of a 10% increase of carbon emissions in 2050 now, so that we can know how much we are willing to pay to avoid any consequences of such an increase. For this reason, we need to be certain now that the model can produce reasonable predictions for conditions that may never actually materialise.

Climate models do not merely exist for their own sake but they are tools that are made for a particular purpose. Their primary function is not only to advance scientific knowledge for the sake of it, but to feed into a public debate over what to do about climate change and to inform political decisions about how to manage carbon emissions. Policy decisions need to be made now, but their impact will only be apparent many decades later, say in 2050. Should the scientific model be of any use in informing policy now, it clearly cannot wait for a falsification or verification until 2050. In conditions where decisions are urgent, but the model predictions are long term, the models themselves are unfalsifiable.

Does that now mean that climate science is not really a science? Maybe it does. After all, climate models lack many of the crucial ingredients of Galileian or Monodian science. What made their approaches so powerful was the ability to build laboratory systems. Galilei and Monod could test specific scenarios and hypotheses in an artificial system that reflected their idealisations. They were also able to systematically vary specific parameters and check the response of the system. This is not feasible in climate science. There is no laboratory system that can be used to test hypotheses about the climate. In a sense, the climate models are the laboratory systems.

Not only is it the case that the models are not verifiable or falsifiable, but different models do not even agree with one another. Far from giving the same prediction, two different models developed by different research groups normally give conflicting results, as do the various models developed within the same research group. The discrepancy can usually be explained by different assumptions about the underlying dynamics. Clearly, at most one of the disagreeing models can be correct.

One wonders what makes these models so different from the predictions of clairvoyants and soothsayers. They do share many characteristics. Soothsayers do not seek to falsify their predictions, climate scientists cannot do it. Ask many soothsayers for a prediction and you will end up with many different answers. Ask climate scientists for a prediction and you will equally end up with many different results. At this point one may be inclined to think that there is no difference between soothsaying and climate science. At least, the traditional virtues of good science, empirical verification, are completely absent from both.

If climate science is indeed no better, why does the government not use clairvoyants instead of climate scientists? Their predictions are equally untested, but at least soothsayers are cheaper. A cynic's response may be that the government already spends billions of pounds and euros and dollars every year on keeping universities and public research institutes going. Given this investment, they may as well use their professors and give them some work to do. At least this gives the appearance of getting value for their tax money.

Climate modelling is guessing—to some extent. There is no doubt about that. However, it is very educated guessing. It is true that the models themselves cannot be experimentally tested, but the information that goes into the models can and has been subject to more or less rigorous scientific assessments. Climate modellers will certainly make every attempt to base their models on the current best understanding about climate processes. It is the nature of computational models that they cannot add new information to the model. All they can do is to tease out conclusions and make inferences from the data that is supplied to the model. Computational models are reasoning machines that can automatically process information. In some sense they are outsourced super-brains that can tirelessly calculate, transform and extrapolate. So, there is one difference to the soothsayers who would base their decisions on the entrails of a dog, rather than the latest knowledge about how air-masses are shifted around the globe.

The problem the modeller faces in practice is that the result of the model may depend and typically does depend on the precise shape and form of the cocktail of information that goes into the model. A feedback missed here and a wrong parameter there may materially affect the eventual outcome. This is where the uncertainty about the model comes in. The underlying factual information about relevant effects on the climate, however, is the same for all modellers. What is different is how this information is weighed up, how it is simplified and how the modeller composes it into a coherent model. This process of selecting the important information and simplifying away the parts that are unimportant is an informal process. There are no fixed rules to guide the process of model construction. The modeller relies on her skills and judgment to decide what to include and what to leave out from the model.

Skills and judgments are difficult to grasp as concepts. The ability to transform knowledge into a working model is not so different from the ability to create works of art. In any case, this is not a task that can itself be left to a computer, but it requires understanding of the problem itself, the objective of the model and an appreciation of the resources that are available. The design of a model depends on the subjective judgement of one (or possibly a few) individuals. This process is not formalised by rules, it cannot be formalised, and the process itself is not tested or testable either. It is something scientists do according to their best knowledge.

This is not to say that the process of model construction is completely subjective, as is the question whether or not somebody likes spinach or not. There clearly are models that everybody in the community of climate researchers will recognise as wrong. However, within the set of models that are controversial, there still is some variation in terms of the predictions that models give. Nobody knows which of those models is correct; indeed, researchers cannot even be sure that any of the currently available models makes correct predictions. We will know in 2050, but until then, climate scientists need to hope, assume and use their general understanding of science and the climate to come up with reasonable models and good quality predictions.

Soothsayers have a range of conflicting predictions as well, but arguably the range is greater. Say a soothsayer reads from the entrails of a cow that the temperature in 2050 will have dropped by 20 degrees compared to now, then this is, by soothsayer standards, a legitimate prediction. Another soothsayer may come to a different conclusion based on her readings of dog entrails. Her conclusion may be that the temperature will have increased by 20 degrees. There is no way to arbiter between these two predictions. In climate models, the differences between different legitimate models tends to be much smaller, because the discretion of the modeller to choose her model inputs is much smaller than that of soothsayers to interpret entrails.

There is another difference. It is true that climate models cannot be tested experimentally, due to their nature, but this does not mean that they cannot be tested at all. Climate scientists use a strategy that is sometimes called *retrodiction*. The idea is very simple. Before the model is used to predict the future, say the climate in 2050, the model is first used to predict the past. They start with the climate in, say, 1950 and then see what the model predicts the climate would be like in 2010. One would expect useful models to pass this test, although passing the test is no guarantee at all that the model would do equally well in the future. The main reason is that retrodiction can only test the model on one single scenario, namely the ones that have actually been the case. These circumstances may not necessarily be the relevant ones for the future. Indeed, typically they will not be. Still, retrodiction can be at least a sanity test for the models, which is better than nothing.

Science? What science.

Popper seemed to offer a good way to distinguish belief in Newton's laws and belief in Vishnu. The difference is that the former can be falsified and most of all is constantly sought to be falsified, whereas with the latter this is not possible. Popper introduced falsification as a general demarcation criterion that separates science from non-science. This has convinced many working scientists.

In reality, many activities that by all standards of reasonableness should be included within science would fail miserably if the falsification criterion was applied. It takes a lot of good-will to shoehorn medical research into Popper's scheme, but even shoehorning fails with climate modelling. This leaves us with two choices. We could either deny scientific status to climate science. This is clearly unreasonable, and serves no purpose. Or we can reject Popper's criterion as irrelevant. This seems the better choice.

The problem now is that we are back to square one. We still do not know what precisely it is that sets the belief in Newton's law or the climate models apart from the belief in Vishnu. We still do not know what this elusive "scientific method" is that supposedly creates such secure knowledge. Perhaps we have to start to entertain the possibility that, maybe, perish the thought, there is no such thing as the scientific method? Maybe it is only a mythical creation like Nessie or Fafnar the dragon.

In many ways it is not surprising that it turns out to be so difficult to pinpoint exactly what science is. There are so many different types of research fields and research problems. Each of them operates under its own assumptions and purposes. Galileian science is interested in idealised systems. Monodian science is a hypothesis driven experimental science. Climate science tries to integrate many sources of knowledge and come up with scenarios for the future. Medical research is all about checking the effects of treatments. All these fields have very different objectives. There is no single scientific method because there is no single science. Researchers from all corners of the academy are united in that they want to find out about things, but the specifics of their interests, objectives and problems are very different from one another. It would have been very surprising had there been a single method that applies across all the diverse fields of science.

Another myth of science is that it is value-free and disinterested. While it is true that the results of science are largely independent of the cultural background of researchers that produce them, the decision to use science rather than soothsayers and witchdoctors is not value-neutral. It is a decision that itself forces certain value decisions onto the user. The example was a health service that is organised along scientific insights and in this way imposed its own ethics on us.

Finally, it is often said that science is valid because it is based on empirical facts. One has to be careful. The entrails of dogs provide empirical information as well, just not relevant ones to predict the future. The issue is not so much to what extent empirical facts are used as a basis of predictions, but how the empirical data is interpreted and processed to arrive at the conclusions. In climate science this step of taking empirical knowledge and transforming it into a predictive model is the weakest link in the process. Even though all the knowledge that flows into the model is validated, the model itself may well make wrong predictions. Just being based on empirical facts is not sufficient. When the model itself cannot be falsified or validated empirically, then the scientist is on thin ice. Popper's method is not applicable. When push comes to shove, the model is the product of the judgment and experience of the scientist. There *is* no scientific method to guide the scientist.

We have now stripped away from science most of the things that one usually associates with it, value-freeness, objectivity, empirical grounding and the scientific method itself. The conclusion must be that there is no magic formula, no demarcation criterion that would allow us to distinguish between science and non-science in a universal and objective manner. There are some science projects that use falsification. There are some that use the Galileian method. Often scientific results are backed up experimentally. There are many examples of science that produce reliable and reproducible predictions. Yet, none of these features is universal. In other words, there is no clear criterion that sets science apart from non-science.

The conclusion we have to draw from this is that there is no rational reason to assume that science produces knowledge that is any better than knowledge produced by other activities. Indeed nobody can pinpoint what precisely it is that makes something scientific as opposed to non-scientific. Nobody knows, in a strict sense, what science is. It must follow from this that we cannot be certain that science, the thing that we do not even understand, produces certain knowledge. We can hope that it does, but we cannot be certain in any way.

Der Meister Tön' und Weisen

When I enrolled at the University of Vienna as a young and hopeful undergraduate student in theoretical physics, one of the first lectures I attended was an introductory talk given by Prof. Herbert Pietschmann. He was the departmental chair at the time. One of the first questions he asked and tried to answer was of very fundamental importance for the new students of the department: "What is physics?" His answer to the question was rather surprising. Instead of trying to give a formal definition he said that physics is "what physicists do."

At first this seemed to me like a very silly answer. For one, physicists do all kinds of things that one would not normally categorise as physics. They eat, they climb mountains, some probably breed dogs, build ponds in their garden or give lectures on Beethoven. These types of activities were not what Prof.

Pietschmann had in mind—he was probably thinking of the research activities of physicists. Neverthless, at the time the definition still seemed silly to me, like defining a quark as an object that is identical to a quark. Meaningless.

But on second thoughts there is another element to Prof. Pietschmann's definition. What he was really trying to say with his definition of a physicist is that there is no formal definition. The field is a very broad one and the problems that physicists look at and how they solve them are changing constantly. A hundred years ago no physicist used computers (because they did not exist) and no physicist was interested in complex systems. Both are established parts of physics today. The only thing you can really say is that physics is the whole breadth of research problems that physicists work on right now, plus all the established knowledge that is taught to physics students. Despite its apparent circularity, Prof. Pietschmann's definition is a very clever way of defining physics, and it can be extended to include the whole of science.

Science is what scientists do. There is no description of science that is more specific than that. If we use Prof. Pietschmann's definition, then a whole new perspective opens up. The question is then no longer what it is about science that makes it the eternal truth, the preferred and privileged knowledge acquisition technique. Science then becomes a social phenomenon, a part of the division of labour in society. Science is what academics and industrial researchers do. Rather than scientists following a set of rules that makes their work scientific, scientists constantly define science implicitly by their actions. What is and is not an allowed approach to problem solving and what does and does not constitute a valid problem is under constant review. There is no global and eternal standard for what science is and what counts as scientific evidence or proof, but these standards are subject to change over time just as the problems physicists are interested in change. Astrology and witchcraft are not part of the spectrum of things that are included right now. This is not so much because astrologers do not follow the scientific method, but because scientists collectively do not feel that astrology is a science. Prof. Pietschmann's definition removes the need to find a well defined criterion to distinguish scientists from soothsayers, lawyers or hairdressers. It is very simple. Scientists do not think of soothsayers or lawyers as being scientists, consequently what they are doing is not science. They are not part of the community because they are not looking at problems that are considered scientific problems.

Defining science as the activity of a particular social group solves the demarcation problem of science, but creates another one: If science is not a particular well defined activity that comes with its own guarantee of rigour and reliability, then we have to ask why we should consider it in any way special. Why should governments turn to science for advice, rather than to the prime minister's grandmother or the village seer? Why should the rules and standards developed by scientists be used as the basis to approve drugs or treatments, rather than those used by the witch-doctor?

Ultimately, we will find, that there are no satisfactory answers to be given to these questions. Probably there are many plausibility arguments as to why scientific advice is better than that of a soothsayer, but ultimately there is no rational way to argue the case for science. The real reason why we use science rather than witchcraft is that scientists have been able to persuade us as a society that their craft is superior. In the free play of forces researchers somehow managed to convince everybody that their activities can be beneficial for state and ruler and they should be funded to do more of what they do. They could offer wealth and prosperity by developing new machines, making new inventions, finding new effects, all of which can ultimately add to the greater good and wellbeing in society. And, of course, they have kept their promise. There would be no nuclear weapons without science, no air-pollution, no global warming and no 24/7 CCTV surveillance on our streets.

The societal respect that science enjoys and the associated funding is largely a result of it being able to convince key-decision makers and the general public at large that it is worth-while investing in science. The contemporary argument involves a narrative that modern economies rely on constant innovation in order to drive economic growth. The basic strategy is to convince the public that if we do not keep increasing our collective skills, then we all end up in sweat-shops producing T-shirts for other, more competitive economies that did keep up their skills.

The good thing about being a scientist is that it puts you into a privileged position to make and win arguments about the usefulness of science. For one, decisions makers themselves have spent their formative years in academic institutions as students. Enough time to infuse the minds of future leaders with academic values. Better still, when decisions are to be made the decision makers commission research. To get best value for money, any advice must be based on scientific principles. So ask the scientists themselves about the value of science. They are of course unlikely to undermine their own professional interests by concluding that science is irrelevant. Whether or not science is indeed important for the knowledge economy, I do not not know. The observation is merely that science has acquired a status in society where scientists can be their own judges. Given that they are merely human, we should not be surprised if they twist the facts a bit in their favour from time to time.

Contradicting the scientific establishment is quite pointless. Anybody outside the circle of science is not a scientist, and does not master the "scientific method." This makes it very difficult for those outsiders to stand up to scientists and be taken seriously by them. By virtue of not being scientists, their arguments are bound to be worse than those made by the members of the priestly class.

If science is that powerful, we should ask who these people are. How does one become a scientist? How does one gain entry? What rules need to be adhered to once you are a member? In the absence of a clear-cut definition of science, some understanding of what science is can be gained by looking at the social structures from which research emerges. Unlike the freemasons or the members of the Austrian People's Party, science is a loose community with no clear defined membership. There is no central control, no pope-equivalent, no formal byelaws or membership register. Still a scientist generally knows when she meets another one, although there are fuzzy cases.

How does one become a scientist? Over time the loose community collectively devised rules for that. The process is a bit like becoming a member of the Meistersingers, the community of guild-organised singers that was depicted so colourfully by Richard Wagner in his masterpiece "Die Meistersinger von Nürnberg." The 15th and 16th centuries was their high watermark, with Meistersinger guilds in many German towns and cities. They had elaborate rules of how to compose and sing songs and also an elaborate internal hierarchy. At the top was the Meistersinger, who mastered the rules and had developed an original song of his own. To become a Meister, one had to progress through the lower ranks first. At the bottom of the heap was the "Schüler," then there was the "Schulfreund," the "Singer," and the last step before Meister was the "Dichter." Admission to the Meister level required the candidate to pass an admission tests. David, an apprentice cobbler and singer describes the process:

Der Dichter, der aus eig'nem Fleisse zu Wort' und Reimen, die er erfand, aus Tönen auch fügt eine neue Weise: der wird als Meistersinger erkannt.

Only the poet who can craft his own tunes and words and shapes them into a new song, will become a Meistersinger. The gruelling judge in this process is the "Merker." He is one of the established Meisters and listens to the admission song of the aspiring Meister, marking off any infringement of the rules of the Meistersinger book. If there are more than seven then the song is deemed a failure and the hopeful apprentice is chased away in shame instead of being given the crown of the Meistersinger. The real Meistersinger existed well into the 19th century, but throughout their centuries-long existence they failed to leave any impact on the music of their days either in Germany or anywhere else. At some point, the Meistersinger just died out because nobody cared about them any longer.

There are probably many reasons why the Meistersinger never made it to social status and prominence. If I had to make a guess, I would say that their rule-book was a big problem. Unlike science they had very specific criteria as to what was and was not a Meistersinger song. Indeed whenever the Merker judged the quality of a candidate Meistersinger, then he was only allowed to judge by this rule-book. There was no arguing with rules. This is probably a German thing, but most of all it meant that the Meistersinger prevented themselves from adapting and developing.

As far as its hierarchical structure is concerned science is similar to the Meistersinger guilds. Professors are at the top and students are somewhere near the bottom. The first step up the hierarchical ladder is very much like the *Freiung* as Wagner describes it. For many the award of a research doctorate is the first point of accession to the ranks of science. Just as the apprentice singer had to demonstrate that he has created a new song, so must the candidate during her doctoral defence show that she has added something novel to science. In addition to novelty any doctoral dissertation has to demonstrate knowledge of the current state of the art in the field and it has to show how the novel contribution fits into a recognised problem.

The Merker in this process is an established scientist,

somebody who knows the field of research very well. There is something crucial that makes the opponent during a doctoral defence very different from Wagner's Merker. She will judge the thesis not based on a set of established rules, but based on her experience and some unwritten conventions. Based on this she will determine whether or not the piece of research is good enough to deserve the degree or not.

This first encounter with opposition during her defence also sets the tone for the rest of the career of any young academic. Within science achievements are always judged by fellow scientists. A new piece of research has to be published, usually as an article in one of the scientific journals. However, before it can be published it has to be subjected to peer review. This means that it is sent to two or three or more other scientists who are invited to comment on it and determine whether the article deserves to be published or not. Again, the reviewers do not follow any rules when they come to a verdict. Instead, they base their decisions on the experience that they have acquired over their professional lives. The same system applies when a scientist applies for money to do her research. She writes a project proposal that is then judged by fellow scientists who will recommend whether or not the money is to be awarded or not. Again no rules. It always comes down to the judgments of the reviewers, what they think is good and what they think is worthwhile and relevant. There is never any need for a reviewer of a PhD thesis or a journal paper to justify her criticism with reference to a set of rules.

This may seem like a strange system. When there are no rules, how can it be guaranteed that the review itself reflects good judgment, approves good science and rejects bad work? The answer is that there is no such guarantee, because there are no rules as to what makes good science or bad science. A piece of research is good enough to be published or to earn the author a doctorate if other scientists think that it is good enough. Good science is what scientists think is good science.

One consequence of this way of organising research is that research fields, once established, have a built in momentum that keeps them going. Say an article in field X is to be published in the Journal of X. Before it can be published it has to be sent for peer review, normally to experts in X. There are a number of things the referee could and will judge, including the technical correctness of the argument, to what extent the manuscript reports novel research, how well it sets the context with existing knowledge in X and so on. This is what referees typically do. What is much rarer is that the referee also asks to what extent research in X is worthwhile in the first place or to what extent anybody outside the field of X would care or be interested in the research. The reviewer is normally herself a member of the community of researchers of X, has published in the field and has built her career or parts of it by doing so. She is unlikely to be inclined to ask these kinds of questions. The importance of X will be self-evident to her. She is completely immersed in the field, she is acquainted with the major players and has over countless cocktails at conferences talked about the future of X.

As long as the field is big enough the chances are that it will continue to thrive. Every professor supervises new research students and hires postdocs in her field. These new blood researchers are socialised into the field of X, learn about its problems and get to know the protagonists and become themselves full members of the community of researchers. Some become professors and propagators of X. When there are many professors interested in a particular topic then they are also more likely to be members of grant panels that allocate money to research groups, further funding the expansion of their field. If a field is growing and attracts funding then it is also likely that hiring committees will appoint new professors within this area. In this war, once an area of research is established it has an innate tendency to grow, because growth serves directly the personal interests of its members. If there is money in a field, then it will attract members. If it attracts members, then there will be more money in the field.

Within the field itself, the question about the value of the field is never seriously asked. In fact, as young researchers immerse themselves into the mores of a subject they are too busy learning to ask the right kinds of questions, and have no time left to worry about the societal impact of what they are doing. By the time they are established researchers themselves, the importance of their field has become nearly self-evident. They are socialised into it and doubting the field just does not come to mind any more. Why the research is done and whether the problems it considers are indeed promising to look at becomes a non-issue.

Research fields are not only areas of enquiry but also fiefdoms of power, foundations of careers. Those who built their career on X and are well established in it have vested interests in the continued growth of the field. While they are a research star in X they are nobodies nearly everywhere else. The experts in a field therefore have nothing to gain from questioning the relevance of their own area of specialisation, though they have much to lose. Outsiders on the other hand are quickly dismissed by the specialists and experts as incompetent amateurs whose views need not be taken into account.

There are few correctives in science once a field is established. Many fields degenerate to knowledge bubbles. Fields that are past their prime continue to grow and attract new researchers, they spew out new results, even though novel ideas may not have been generated for decades. Anything new is now only an improvement to a minor variation of an improvement of another variant of an idea that had been great 30 years before. Senescent fields are sometimes hard to kill. Too much is at stake for the individual researchers. They may not even be aware that their field is dead and keep filling the journals with pointless articles, hire new researchers to do anaemic research and shuffle research money to feed the vicious circle. By any standard this is a huge waste of resources. Somebody should do something about it and clamp down on these bubbles. Then again who would that person be? Whose right is it to decide that a field is finished. Maybe the revolutionary new idea is just around the corner. Perhaps the solution of the grand problem is about to be discovered. Just like asset bubbles in economics, it is difficult to be certain that a knowledge bubble is indeed a bubble and not just a normal rough patch of a field that is about to renew itself. Science is what scientists do, and good science is what scientists think is good science, and even within a bubble there are enough of those who think that the research in their bubble is great.

All this is not to say that there are no correctives at all. Eventually a research field that fails to inspire suffers the same fate as the Meistersinger guild. It becomes marginalised and loses its appeal. The professors will find it harder to attract students to the field and postdocs change their research interests if they can. Also, many academics are in several communities at the same time. Membership of X does not preclude simultaneous membership in Y as well. If Y is more exciting than X then many members will start to attend the conferences in Y more than in X. Probably they stop their activities in X altogether and the field of X has lost another member. So, in practice fields do die, but often they can live well past their prime.

This shows that science is a social activity. Scientists are humans and within science they show the same kinds of patterns and behaviours that they show in other contexts. Like society at large, science has its own fashions and fads. Huge sun-glasses are fashionable one year, miniskirts the next. One year the community of researchers takes a liking to a particular type of research and a huge swell in activity ensues. A few years later what was once fashionable may have become a knowledge bubble and nearly disreputable. Everybody who has been in science for a few years has seen the ups and downs of fads, the comings and goings of sexy research topics.

A strong driver of science and scientific development is the competition between scientists. Amongst other things this is a competition for ideas and print space in the best journals. As in any community, it happens over and over again that there are differences in opinion, different schools of thought, different approaches to one particular problem. Sometimes these differences can lead to heated discussions, but often the differences are milder and there is a rational exchange of ideas and arguments between scientists. There is every incentive for one scientist to find flaws in the work of others. Any criticism will be another research paper, and probably a good one too, and as such it will advance the status and career of the author. There is absolutely no guarantee that the bones of contentions are real issues, or of any ultimate relevance, but there is a guarantee that whatever the idea it will be scrutinised according to the standards of the particular field of research. This does not lead to eternal truths, but it means that arguments are evaluated thoroughly. Science at its best is like a murder trial but defendants are only convicted if there is evidence beyond any doubt (reasonable or otherwise). (Although they are held in jail until proven innocent!) A good example is the work by Müller and Peters (see chapter 2). There has been so much evidence that Einstein's theory of relativity is correct that one would be inclined to simply accept it. However, scientifically there is still some benefit in systematically doubting it.

Despite all this scrutiny, the internal debates within a field are heavily framed by the collective interest of the community and community standards that have developed historically. What precisely these standards are is nowhere written down. Still for the practitioners of a field they have become second nature, and they are acutely felt by the student trying to get into the field or a scientist who switches. In mathematics arguments are always formal proofs of theorems. The standards of rigour are high. In physics arguments are typically mathematical as well, but rather than water-tight proofs physicists are often content with derivations, a less rigorous form of mathematical argument. Then of course in physics there is also the experiment. Climate scientists argue from a mix of experimental data, physical insight and computational modelling. In micro-biology arguments are nearly exclusively based on the verbal interpretations of experimental evidence in the light of hypotheses. Biologists are usually unimpressed by mathematical arguments, just as mathematicians have little sense for the experiment as a source of knowledge, at least when it comes to mathematical knowledge. Despite the differences, within the historical parameters of each field that determine what an argument is, permanent scrutiny is the nature of the game. Constant competition within the community keeps everybody on top of her game. It is hard to get away with sloppy arguments.

Science is a social game. It is a competitive economy of ideas and arguments. Every member of that system tries to compete by doing something new, finding an argument that nobody has made before, coming to a conclusion that has not been drawn or even finding a phenomenon that is undiscovered. Being a competitive system, there is also self-scrutiny built into it, in that members review and systematically doubt each others' results. Science is also a social system with fashions and fads. When an idea is right for its time it gets picked up by many and maybe develops into a field. When the time for an idea has not come yet, it may be ignored. What is researched today and what will be researched tomorrow depends on where the community is going collectively. Predicting what science will be interested in in 20 years is as difficult as to predict what types of clothes people will wear in the future.

The unspeakable

Science is a social activity. Just as bin-men, hairdressers, lawyers and soldiers and many other professions have developed their own codes of behaviour, so have scientists. Science is a social process, a fluid network of people who feel that they belong together and who see themselves as part of the same tradition; science is a loose community that shares some values and that has some common understanding of itself, what it means to be a scientist, what sorts of things a scientist does and what her place in society is. Procedures have been devised to regulate who can be admitted into the ranks of science and what one has to do to be recognised. Internal reward systems have emerged. Academia also has its own pecking order. People within the system are normally quite aware as to where they are on the ladder of achievement. If you work at one of the top universities you are probably higher up than somebody who is somewhere in the lower tier. If you are a professor you tend to be higher up than a graduate student. The system is complicated, but most people within know where they are and what it takes to get up.

Scientists are also the ones who make the rules about their own standards. They decide what makes good science just as lawyers collectively decide what makes a good legal argument. Scientists in the making are socialised into the standards during long years of intense training. Once they become professors themselves, they have a stake in keeping the standards. Once you have been in a system for a long time, you no longer see it with a critical eye. The unwritten rules that were so hard to learn for the aspiring young doctoral student have become second nature. This guarantees a certain stability of the system against change. However, science is not immutable. Things do develop as time goes on and standards are changing. Science is not a homogenous rigid structure either. The traditions in life sciences are slightly different from those in, say, geology, which are again somewhat different from computer science or physics.

It is not at all surprising that nobody has been able to give a non-circular definition of science. Social processes, science included, are nearly always very complex. With every article that is published, the referees and the editor of the journal makes a statement about what they think good science is. Similarly, with every paper that is rejected, they make a statement as to what is not good science. It is quite hopeless to try to shoehorn this complex social system of constant negotiation into a simple set of rules.

There is no rule-book for science. To be successful, science has to be flexible. The Austrian philosopher of science Paul Feyerabend famously summarised this when he said that "anything goes." He probably exaggerated a bit. Many things "go" in science, but there are still no-no's that will not go. There are some very fundamental assumptions in science that will never be violated, some rules of engagement that are kept by everybody, be it an elementary particle physicist, an economist, historian of film or a scholar in old Egyptian languages: For any result of research, it must be possible to communicate the contents of this result to other scientists. It must be possible to articulate the knowledge in words or to write it down. This is non-negotiable.

This requirement to produce inter-subjective knowledge sets research apart from many other activities in life where knowledge can be both subjective and still meaningful. For example, young lovers do not need to articulate their feelings for one another. They will be happy with an emotional understanding of their feelings. The followers of a religion will often have the same experience. They can be in direct communication with their God, but their experience will not have an empirical trace. It will not be possible to make the personal revelation public.

Much of our life is about perceiving our internal worlds. Some aspects of these internal experiences can be shared, many cannot. These purely internal experiences are excluded from meaningful scientific discourses. These unspeakables are limits to science. Science can only be about things that can be put into words. When something cannot be articulated, cannot be shared, then it cannot be part of science. This is a basic rule of research. It could not be any different. The unspeakable, the inner feeling, cannot be written down, it cannot be articulated. Other researchers cannot scrutinise it, so it cannot be researched.

For example, Synnøve thinks that young kittens are cuter than puppies and Lars prefers to spend his Easter vacation fishing in the sea rather than skiing on the mountains. Such personal choices are realities. Some people simply do not like dogs, even puppies, and some people hate skiing. Synnøve and Lars' preferences will not be the centre of scientific discussion. It is not possible to argue about taste, indeed the preferences of those two particular people are not even of interest to anybody except themselves and probably their immediate social environment. However, while the individual tastes of these particular two people are not accessible to scientific enquiry, the physiological processes that correlate with these preferences are. Preferences presumably correlate to events in the brain, which can be measured, modelled and studied. It is a meaningful question to ask scientists whether people who like fishing have slightly different brain patterns than people who like skiing or whether the preference for cats over dogs is somehow physiologically reflected. This is, however, not the same as asking why Lars has the particular preference he has. If we asked him why he likes fishing more than skiing he would probably just shrug his shoulder, or he would say that his parents forced him to ski when he was young and he dislikes it now, or he simply thinks that it is too cold, or too dangerous or he likes it, but probably less so than fishing. As far as his choices and preferences are concerned, the relevant reality is the internal experience, the particular feelings he has during skiing and fishing. In any case, if Lars explained his particular preferences he would not point to neural correlates of this internal experiences. What is more, if a scientists explained to him that he likes fishing more than skiing because of his brain patterns, Lars would most likely find it difficult to relate to such an explanation. It would not be a relevant explanation for him.

Another example: Is it a scientific question whether or not the Mona Lisa is a good painting? No it is not! The judgment about the quality of the artwork ultimately is one that comes down to taste. For some reason the Mona Lisa has managed to capture the imagination of those who have seen it. It exerts a magic, it has something mysterious about it, some quality that makes it the most famous painting of all paintings. Its effect on popular culture and imagination can be studied and described by sociologists, but this does not explain why there are so many people who are fascinated by the painting. Neuroscientists can perhaps one day understand what is going on in the brain of those who admire the painting. Physicists and historians of art can use sophisticated techniques to study who it was that da Vinci painted in the Mona Lisa, what strokes he used, how many layers he applied to the painting and so on. However, the mystery of the painting itself, why John Smith likes it, but Ian Miller does not, is down to individual taste, which is not explained by neuroscientists, the structure of the painting nor by the sociologists' accounts of the role of the Mona Lisa in popular culture. Just as Lars' preference for fishing is beyond the scientific discourse, so are John and Ian's tastes regarding the Mona Lisa.

The difficulty for science in accessing personal tastes and preferences is not so much the lack of knowledge, but the problem of accessing the data. When you eat a piece of seafood, then the delight or the disgust (depending on your taste) can only be experienced by yourself. You can communicate to the environment that you like or dislike it, but the sensation itself is private to you. A shrimp in someone's mouth triggers a very particular experience that is pleasant for some, for others it is not. This sensation must remain private. It is in principle inaccessible to the scrutiny of others and therefore inaccessible to science itself.

When I ask whether your sensation of eating shrimp and that of somebody else are the same, I can do so without violating any grammatical rules of language. Still, the sentence is meaningless. I have no way of knowing what shrimp eating feels like to you. Maybe for seafood lovers shrimp tastes more like Wienerschnitzel tastes for seafood haters. Maybe seafood lovers have exactly the same sensation as seafood haters, but they somehow learned to interpret it differently. One thing is certain, however: We will never know. Whether or not two people feel and perceive the same things when they experience flavours sounds like a meaningful question to ask. On closer inspection it turns out that it is not.

To see this consider how one could go about comparing the perceptions of two tasters? One possibility is to compare the brain activity while eating shrimp and infer from that whether or not the experiences themselves are the same. This is probably not very interesting and certainly misses the point. A scientist may be able to infer from a record of the brain activity during shrimp-eating whether or not the subject enjoyed the activity or not. A brain activity record can show what part of the brain has been active and how active it has been. At the moment such records are quite coarse and summarise the activity of a large number of individual neurons. It is not inconceivable that in the future brain scans could reach resolutions that allow researchers to observe brain activity down to the level of the individual neurons. If science got that far one day, and it is certainly conceivable that it will, then it would be fair to say that extremely detailed activity scans of this sort can capture all the information there is to be gained about the brain. A clever scientist may even be able to tell from the brain activity that you were eating shrimp just by by looking at the particular curves.

However, no matter what detail the brain-record contains, being able to tell that somebody eats shrimp and being able to model and describe the neurological processes that are going on while she has the experience is not the same as reproducing the experience itself. The neural correlates of these experiences are irrelevant. Even the best and most accurate brain-scans would not induce a taste of shrimp on the scientist's tongue. She could not feel what the subject felt when the scan was taken. No matter what amount of detail the record contains, no matter how sophisticated the brain-reading and brain imaging devices become, it will never be possible for anybody other than the subject itself to experience the sensation of shrimp on her tongue.

One could object that it is possible to describe tastes. Experts of wines often describe flavours by comparing them to fruits or butter or oak, which then allows other wine experts to learn about the particular wine. This is not a counter-argument. Describing the flavour of wine in terms of other tastes does not communicate the sensation itself, but only describes it in terms of other sensations. The reason this description makes sense to others is that they themselves have previously experienced the tastes of fruits and butter and oak for that matter. In this sense the descriptions of wine are just very imprecise ways of pointing to supposedly shared experiences.

To clarify, imagine a scientist who suffers from a genetic condition that means that she never had the ability to taste anything. Descriptions of the taste of wine or the taste of shrimp would be quite meaningless to her. They would not re-create the experience of biting into a shrimp or the feeling of the flavours as they develop in her mouth. Despite her impediment she would still be quite capable of interpreting brain-scans. Nothing would prevent her from understanding that somebody else just enjoyed/loathed shrimp given a brain scan. This tells us that the objective observation (brain scan) is very different from the subjective experience (shrimp taste).

Let us now come back to the original question. Can two people determine that they have exactly the same subjective experience when they eat shrimp? There are two conceivable interpretations of this question. One can say that two people have the same experience if their brain scans are somehow equivalent during the experience (they will never be exactly identical, but they would be sufficiently similar to count as identical). Testing for equivalence in this sense is fully possible and indeed an interesting and feasible scientific question. The second way to understand this question is to ask whether the underlying experience is the same. Are two people who eat shrimp really experiencing the same sensations? This second question is not reducible to the first question, just as Lars' dislikes cannot be explained by brain-states and just as the detailed descriptions of brain scans of shrimp-eaters cannot induce a taste of shrimp. This second question is also meaningless because we have no way to convey our inner experiences except by relating them to other inner experiences and assume that they are shared. There is no way in which inherently subjective realities can be made objective.

Sensations like tastes, preferences, feelings and experiences are beyond the reach of science. This does not mean that their physiological correlates cannot be studied, but the experiences themselves are not accessible to a public discourse. They are private to the individual. This means that they cannot be critically explored by other scientists and therefore they just do not fit into the conventions of scientific exchange. Science requires that the objects it studies are accessible to everyone, at least in principle. There is no way, however, that I could order a sample of Lars' feelings about skiing from the scientists who studied him. Science requires inter-subjective data.

That experiences are not accessible to science does not, however, mean that these experiences are not real. Quite clearly, our subjective impressions of the world are perhaps the strongest aspect of reality that we can imagine. They make up an irreplaceable part of our lives. Diseases that take these experiences away are tragic events. A blind person's sight cannot be replaced by descriptions of the visual world, no matter how accurate the description. The loss of taste cannot be compensated by words.

It never fails to surprise me that very eminent scientists forget about the distinction between the unspeakable and science. In public debates about science it is all too often forgotten that science can never be about our inner life. The territory of research ends where the kingdom of the unspeakable starts. Arguably, most of the things in our lives are unspeakables. When we want to express those, then we need to use different modes of communication to be effective. A science of the unspeakable is meaningless, but that does not mean that the unspeakable is meaningless. It just means that there are inherent limits to science.

Science and magic

In their last performance Siegfried and Roy worked again with the tiger that years before had ended the career of two of the most famous magicians by seriously injuring Roy. Roy, now barely walking, led Siegfried into a transparent cage. The curtain comes down over the cage, just for a moment. The curtain opens again and the tiger is in the cage. A moment later at the other side of the stage a second cage is revealed and Siegfried is in there. Siegfried and Roy and the Tiger say good-bye to the audience at the end of their last performance.

Magic shows like this work for two reasons. Firstly, they are all about good showmanship. Secondly, they show events that apparently break with what we normally expect to see in our environment, the laws of nature. This is the magic part. In our daily lives we do not expect to see a man being instantaneously replaced by a tiger, nor would we expect a man to be transported over meters within the blink of an eye. Yet, in Siegfried and Roy's show it happens. Magically. Or apparently so.

The audience knows very well that there is no real magic in

this. Everything that happens is fully consistent with the laws of nature. Whatever seems to be magic, be it a man replaced by a tiger or a woman being sawed into pieces just to be put back together again, must clearly be an illusion, a trick. What seems incredible has often quite mundane explanations once we see the things the illusionists hide from us with skilled misdirections. Even the most naive member of the audience knows that it *is* a trick not sorcery. In principle everybody can perform the trick. No special powers are required. Part of the appeal of a good magic show is the guesswork after the show. How has he done the trick? How could Siegfried switch place so fast? How did they saw the woman in half?

Derren Brown is a British illusionist specialising in tricks with a psychological twist. He seemingly reads peoples' minds or anticipates participants' thoughts, contacts the dead, pretends to be a medium or fills people with the holy spirit. While all this is shown, the TV audience at home is always reminded that Brown does not actually claim to have any special powers. All he is doing has natural explanations. Brown never fails to make the audience aware of the educational point: If he, without any special spiritual powers, can do what the self-proclaimed media or clairvoyants or commanders of the spirit do, then maybe these latter ones are charlatans. At least, there is no need to believe in anything paranormal to explain what they are doing because people without any specific spiritual skills can do the same things.

These programs make fascinating viewing, certainly great entertainment, but Brown, who lives as an open atheist, has an agenda. If illusionists are good at appearing to transcend the physical laws, dear audience, be careful next time you come across somebody who actually claims to have some paranormal powers. The implied message is an anti-religious one. If you see miracles, be skeptical, as it may not actually come from God, but just be a simple trick of the mind.

When you watch Brown's shows, you will notice that he likes to sprinkle science into the program. Often he refers to specific psychological experiments. Science in general is also often mentioned, usually in the sense of it being a force of reason. Science as the antithesis to miracle and belief. As the audience, we are constantly reminded that everything that happens in his shows has natural, that is scientific, explanations and no miracles are required for any of his tricks. The laws of nature are never violated. (What a relief!).

There is, of course, nothing wrong with trying to expose crooks and pretend-spiritualists. However, some caution is warranted. First of all, being able to give a naturalistic explanation for a supposedly super-natural event does not prove by itself that the event to be explained did not actually involve miracles or magic or in some other way violated the laws of nature. For example, Siegfried and Roy's trick has a naturalistic explanation, in the sense that they could tell us how to replicate the apparently instantaneous transport of Roy and the Tiger between cages. However, this does not logically preclude the possibility that in the future somebody will repeat the act using real magic. Admittedly, this is somewhat far-fetched, but it should be borne in mind.

Moreover, just because many or all supposedly para-normal or magical tricks can be accounted for by naturalistic explanations does not mean that no real magic has been used to produce them. It could well be that in the future a real magician performs some tricks that involve genuinely super-natural powers. The problem with those real super-natural events, if they really indeed exist, is that they would be very difficult to recognise as such.

In order to illustrate this, assume that Derren Brown finds his master. Somebody who is an even better magician and can do things that Brown cannot do. Consequently, Brown could not assure us that all the tricks of the super-Brown happened in accordance with the laws of nature, because he does not know how they are done. Derren Brown is good, but sooner or later a better magician is bound to come up.

How about us, the audience? Presumably, we would be quite amazed by super-Brown's tricks, scratch our heads wondering how the tricks work, but not for a moment would we think that she really has any super-natural powers. Super-Brown is just a very good illusionist. What, now, if super-Brown is in reality an evil version of an illusionist and claims that she has magical powers and is not a mere trickster. We would all be sceptical to say the least. After having gone through countless myth debunking sessions with Derren Brown on TV we are not inclined to believe that anybody has super-natural abilities. We could not help but assume that she is a trickster (or rather fraudster). Most likely, we would be right.

However, there is of course another possibility. What if the super-Brown was in indeed not a super-Brown at all, but truly had para-normal or magical powers. Most would find that quite outrageous as a thought, but who knows. Maybe these people exist. In any case, let us assume for the sake of the argument that super-Brown has indeed true magic powers. Who would believe her? I certainly would not. There have been too many great illusionists who have done apparently impossible things and none of them was a real magician. Even if her super-natural acts looked incredible, I would be confident that there is a naturalistic explanation for them. Indeed, the more incredible the deed, the more sceptical I would get. I suspect that my scepticism would be shared by most people.

An interesting question is now the following: What precisely would super-Brown have to do to convince me or her sceptics of her real magical powers. What if she walked on water, or healed the sick? What if she woke the dead? What if she made me fly through the air? I suspect that I would remain sceptical no matter what. Believing in miracles is not something we do these days. It is not in our zeitgeist to think in terms of magical powers. On the other hand, it probably would not have taken much to convince a medieval farmer of super-natural abilities. Maybe a modern hand-mixer or a mobile phone would have been enough to make him believe that you are a God.

Distinguishing super-natural events from events that can be explained in proper naturalistic terms is more difficult than is first apparent. Sir Arthur Clarke was a science fiction author. Apart from his legacy as a writer he also left to the world his three laws, the third of which says the following:

Any sufficiently advanced technology is indistinguishable from magic.

Imagine that it was indeed not a trick that transported Siegfried from one cage to the other, but a Star Trek like teleportation. Or imagine, that Derren Brown does not use trickery to read peoples' minds, but he has access to some very advanced brain scanning, that allows him to infer what people are thinking about. As mere onlookers, we would not be able to tell the difference between normal illusions, real magic and advanced technology. In that sense, technology is indistinguishable from magic. In much the same way Clarke's law also applies to natural science, the unlocking of nature's advanced technology if you will. We are used to seeing bees fly and worms crawl, we are used to seeing people walk around, babies develop from apparently nothing. Most of the time, we do not think so much about it, because these sights and processes are so familiar to us. If we take a step back and reflect on it, then we see that these everyday events are quite astonishing. What is it that allows a worm to crawl, a bee to fly or cells to assemble into a baby?

We do not know, but we are trying to find out, that is, science is trying to find out. And it is doing so quite successfully. Every year, every month new things are learned about how nature works, what makes worms crawl, bees fly and babies develop. To our ancestors all this may have appeared to be magic. The unexplainable act of some superior beings. Consequently, life was attributed to some vital force, thunders were the rage of the gods and so on. So far science has not cracked all of the mysteries, but many of them. Phenomena that seemed unexplainable manifestations of the mystical now have very dry scientific explanations. There are still gaps though, things that we cannot understand. However, just as we are inclined to believe that super-Brown is just an illusionist, we are also inclined to believe that all the tricks of nature that we have not cracked yet will be revealed eventually.

For example, science does not yet understand what life is, neither does it understand how life originated. There are some theories, but it is safe to say that none of those is convincing…yet. Scientists have the firm belief that it is possible to explain life and its origins in terms of physical and chemical interactions only and they do not assume for the moment that it was Vishnu's powerful hand that created life. The fact that we do not know right now how to explain it, does not mean that the phenomenon is in any way mystical or unexplainable or requires magic. Since organisms are physical processes, there must also be a physical explanation for them. Life may still look like magic, but it just is an example of nature's very advanced technology.

Scientists are very much like the audience of a magical show. They admire the wonders of nature, but then they try to figure out how the trick is done. This desire to discover the trick is the key driver in science. It would make no sense for a scientist to simply shrug her shoulders and admit that she will never find out, or that some observation she made is unexplainable or requires some supernatural agency. Invoking a divine creator is not a conceivable part of a scientific theory, no matter how unexplainable an event appears to be. The assumption always is, has to be, that nature is an illusionist, a trickster, not a real magician.

A corollary of this is that it would make no scientific sense to assume that the world is created by God. This would not answer any questions. So, even if somebody came up with irrefutable evidence that some sort of creator God indeed existed this would not solve any questions. On the contrary, it would open up a whole new field of research into the nature of God. How did she do it? What is this God made of? How did she come about in the first place and what are her properties. Who created God? Why is there not "no God" rather than a God? What are the particular mechanisms that allowed her to create the universe? Since she existed before our universe was created, she must be living herself in some other type of proto-universe. What are the laws of this proto-universe? Is there maybe another God-God that created our God? Is God indeed a woman?

Explaining any effect in terms of higher powers is not an explanation at all. It merely shifts the problem from one level to another and does not solve anything. The idea that nature is anything but a grand illusionist is inconceivable from a scientific point of view. The idea of events that are not consistent with knowable laws does not make sense within science. There is no way scientific research can possibly find out about super-natural beings because this contradicts the built-in assumption that nature is a trickster.

Even if there were indeed super-natural events, say miraculous healings, or the dead coming back to life or men walking on water—as long as these events are very rare or even singular events a proper scientific study is not possible. Such events would not attract the attention of serious scientists. If they occurred predictably or there was reliable documentation that allowed systematic study, then it may still be ignored, or it is researched until natural explanations are found. Up to that moment the phenomenon will remain unexplained. At no point will it ever happen that science declares a phenomenon a miracle or due to super-natural agency, because this would not be science any longer.

Chapter 6

Scientific rationalism

Mein Schlaf ist Träumen, mein Träumen Sinnen, mein Sinnen Walten des Wissens
Siegfried, Act 3

Optimism and hope

We live in a time of general scientific and technological optimism. There is an underlying assumption that things are getting better all the time. Computers are getting faster, cars safer and cleaner, medical treatments more effective. More generally, we do believe that there is a technological fix for every problem. There is a great intellectual optimism. The idea that we will be able to reveal all secrets of nature. Eventually we will have a theory of everything, we will be able to understand the brain, we can predict and control every system and achieve "total" scientific insights. For sure, we are not there yet but we have figured out how to get there. There are no limits to knowledge any longer. This is the belief.

For sure, this optimism is not shared by every single member of society. There are some who do not believe in scientific progress. There are also a few who are not convinced that scientific progress always translates into progress for humanity. Notwithstanding those sceptics, there seems to be a broad consensus in society that science is a good thing, will improve our lives and should be supported. The optimism is understandable. Life conditions, wealth, efficiency, convenience have improved very much, largely as a result of scientific developments. Even within our lifetimes science and technology driven changes, mostly improvements are tangible. Science has also contributed to our cultural and intellectual life by producing a constant stream of insights into nature, which feed back into the way we define ourselves. We now know things, understand systems, that we previously believed incomprehensible. The question is: Does that mean that it will continue like this forever? Is it rational to be so optimistic? Is it rational to believe that scientific progress will continue forever?

Thinking about this for a moment, it is easy to see that there are no rational reasons to assume that science will indeed continue to provide the benefits it has provided in the past. There is no reason to assume that the scientific discovery will never come to an end. It could well be that one day technological innovation will stop. The fact that science has uncovered many secrets of nature and that our ability to do things and our economic and physical wellbeing has improved so much over the last two centuries does not mean that this trend will continue. It is not a law of nature, let alone a logical necessity, that things will get better all the time. They may well stay the same or become worse.

Even if scientific discoveries continue to be generated the day may well come when scientific thinking does not provide the right kind of reasoning to solve problems any more. At some point it could be that that different kinds of enquiries, different kinds of knowledge are required to solve the problems of the day. Indeed, this probably is already the case. Scientific enquiry is often used in the context of decision making and governance. Decisions about the construction of new power stations, genetically modified food, medical treatments and so on regularly involve scientific research as input to get an idea of their safety, sustainability, efficiency and so on. However, the ultimate decision whether or not to build the plant, pass the law or approve the treatment does not depend on the scientific data alone, but often involves other aspects as well.

Scientific assessments of the safety of, say, a nuclear power station are based on a number of assumptions. Whether or not these assumptions are true is often uncertain. For a nuclear power plant one would typically want to be confident that the geology in the area is robust, that there are no major earthquakes that could threaten the plant. Or, if there are earthquakes, one would want to be certain that the structure can withstand the earthquakes that are likely to occur. A problem is that there is no maximum strength earthquake, but only a distribution of strengths. So, one can say that an earthquake of strength x will occur with a certain probability within the next 10 years. When it comes to building the power-plant one has to think about what sorts of probabilities for catastrophic destruction one is willing to accept bearing in mind that reducing the disaster probability increases the cost of the project. Is 1 in ten thousand low enough, or 1 in ten billion? Ultimately there is no scientific answer to that, but a societal consensus has to be reached as to what level of risk is acceptable.

It is difficult to properly foresee cataclysmic events in a complex world. In order for a scientist to make any useful inferences about the world, she needs to systematically disregard connections between her system of interest (the power plant for example) and the environment. In a way this is similar to the Galileian idealisation which creates systems by systematically ignoring interactions. There is a crucial difference though. Galilei could be completely certain about the validity of his particular system in two ways. Firstly, he had experimental access to the system. He could try out how bodies fall. This no longer works in real life cases. Secondly, if he got it wrong, it did not matter. Except for his reputation as a scientist there was nothing at stake. Galilei did not have to worry about catastrophic consequences for thousands of people if he made a mistake. Risks associated with nuclear power plants and other real life systems are very different in this respect. They cannot be checked experimentally and errors can have fatal consequences.

Also problematic are unintended consequences. In real life,

systems are rarely isolated from one another. Everything interacts with everything else, although often these interactions may not be that obvious. It is practically impossible to incorporate everything into a model. Neither is this necessary, because most interactions are weak anyway. Taking them into account would overload the model and altogether reduce the predictive accuracy, rather than increase it. Though the problem is that it is not always clear what is and is not an important interaction. Seemingly innocuous effects can become important under certain circumstances, which can be very difficult to predict.

One near classical example that illustrates this well is the introduction of the Nile perch into Lake Victoria. A brilliant account of this is given by Tijs Goldschmidt in his book "Darwin's Dreampond." Lake Victoria is one of the largest fresh water reservoirs in the world and is shared between Tanzania, Uganda and Kenya. Geologically, the lake is rather young but it still had a diverse ecology of small cichlid fishes. Genetically these fish were very similar to one another but they inhabited very different ecological niches. Apparently they had fanned out from some common ancestor over a short period of time. While these species were interesting for biologists to study evolution in action, they were not that great as food. Small and bony, they are ok to eat but by no means a delicacy. Partially for this reason, the British—then the colonial power—introduced the Nile Perch. This is a large fish that is fun to catch and good to eat.

The intended consequence was that Lake Victoria would finally have some good fish to catch. The problem was that meddling with eco-systems is always a risky business. The Nile perch can be quite a fearsome predator. The cichlid fishes had no immediate defence strategies. Predictably, at first, the new arrival started to eat them, driving them into extinction, or at least to the edge of it. This should not have come as a surprise. Much less predictable were some of the side effects, the unintended consequences. One of them was an explosive increase of mosquitos in the lake area. The original cichlids, now severely reduced in number, used to prey on the mosquito larvae. When the cat's away, the mice will play, or number of cichlids down means number of insects up. Another change was economical. Cichlid fish used to be caught for local consumption predominantly for the economies around the lake. Nile perch being a more valuable fish is also more interesting for the export market. Hence, fish is caught, packaged and shipped away. This brings jobs for some, but makes fishing harder for those who are not employed. One scenario, however, did not materialise. The fear all along was that the Nile perch would first drive the cichlids into extinction before getting extinct itself. Luckily, this never happened. There was nonetheless a marked shift in the ecology of the lake.

The Nile Perch story is just one example of how the manipulation of a part of a real world system (introducing a new species of fish) can have quite unpredictable consequences (mosquitoes, change of economy). The Lake Victoria example is not an exception. It describes a pattern of unpredictability that is encountered over and over again. When we interfere with real systems, we know that this can have unintended consequences. What we never know is what these unintended consequences may be. Scientific methods can maximise the predictability, but science is not able to fully predict the consequences of a nuclear power station, the introduction of a new species of fish or the use of genetically modified crops. It is not really surprising that science deals poorly with unintended consequences. Science is at its best in physics, when it is about laboratory systems, when the research question can be focused on a narrowly defined problem, an idealised system that can be experimentally approximated. This is not the case in real world applications of science.

Science with its intellectual roots in the idealised systems of physics does not deal very well with real world cases. This is not surprising because Galilei's big break was precisely that he started to liberate himself from the constraints of the real world. He was no longer interested in what was going on in reality, but what would happen in an ideal system. This eliminates all uncertainty about unknown interactions from the system. The Galileian method works poorly when it is applied to real world complex systems, such as Lake Victoria.

The Galileian approach is not well suited to predicting what is going to happen in real systems because there are no well defined boundaries to real systems. In the Lake Victoria case, it would not be clear what a scientific model should include and leave out. Maybe the mosquitoes are part of the model. Maybe not. Maybe the economic system around the lake should be included and how it may act back on the lake itself. Prospectively, it is very difficult to know which interactions are relevant and which ones are not, and hence by extension, which effects need to be included and which ones can be left out. Moreover, even if it were known what is and is not relevant, it is difficult to know precisely how strong interactions are. In other words, there is considerable uncertainty about real systems. In laboratory systems this sort of uncertainty is removed during the process of building the laboratory system, by polishing the balls and lining the ridges with parchment or by growing the bacteria on a well defined medium with a known amount of lactose. When science comes out of the lab, this is no longer satisfactory. Earthquakes do not care about the assumptions of the scientific advisors.

This implies that scientific reasoning, sometimes, has to be supplemented by other considerations. Large scale interventions into real systems, policies and decisions that affect hundreds, probably thousands or even millions of people cannot be reduced to a mere scientific question. Often there will be room for science to add perspective, but the scientific perspective itself must be scrutinised. It is crucial to understand what types of assumptions the science is based on and what the uncertainties are that the science may hide. Scrutinising scientific input and its quality is then itself a non-scientific process, almost by definition.

Is science the only rational thing to do?

These days many people think that scientific reasoning is the only correct way of reasoning. They see science not only as an effective way to generate knowledge but claim that it is the *only* source of valid knowledge. What else could there be except for science? What other reasoning should there be? Was it not the achievement of science to get rid of all the superstition and false thinking that had led to witch-hunts, sacrifices to Gods that do not exist, totalitarianism and all other vices in the world? Was it not science that freed us from fatalism and myth, liberated us from the servitude to irrationality and enabled us to become free-thinking, autonomous individuals?

It is very hard to deny the positive influence that science and scientific thinking has had on our modern societies. Much of what we take for granted would not exist had it not been the enormous practical and intellectual influence of science. However, we must not fall into a *zeitgeist* myopia. While scientific rationality is a good thing, it ceases to be so if it is universally adopted and becomes the only admissible form of reasoning. Unfortunately, among some groups of society there are detectable tendencies in precisely this direction.

Over the last decades or so a school of thought formed among scientists but also lay people who feel the need to fight non-scientific or even antiscientific thinking in society. This school of thought is very often associated with an extreme, almost evangelical form of atheism. Proponents of this school of thought publicly denounce, beliefs, decisions, forms of reasoning that are not in accordance with scientific ways of thinking. The targets of their campaigns are first and foremost all sorts of religions and religious practices, but also mystics, para-psychologists, non-traditional forms of medicine and so on. I would like to call this school of thought *Scientific Rationalism* (SR). SR is the idea that all there is can be described by science, if not by present science, then by future science. If it cannot be captured by either present or future science then it does not exist or is irrelevant.

SR is a strange idea. On the one hand, it is quite evident that there are many types of knowledge, many things in our lives that are non-scientific, but still reasonable and rational. Obvious candidates are personal feelings, relationships between people, intuitions and so on. As we have seen above, there are also occasions in public life where non-scientific reasoning should be applied because of inherent limitations of science itself. So, it seems that SR is not a very promising idea.

On the other hand, science is an intellectual and cultural influence on all of us, scientist and layperson alike. At least when they are well presented scientific arguments can be convincing to the point where it feels impossible to contradict them. This is not surprising. After all, research takes place in an arena of ideas where opposing opinions are constantly tested in a fierce battle of arguments. The weapons are formal models, empirical measurements and simple persuasion. Whatever survives these gruelling battles is proven in combat. This gives the specialist researcher an intellectual supremacy within her domain and makes it very hard for those who are not scientifically trained to engage in a fair argument.

Scientists have seen an opportunity here and have not been lazy to grab it. Scientific thinking has seeped into popular consciousness through countless talks, books and TV programs. The woman on the street invariably knows about the magic of the scientific method, this particular way of doing things that allegedly creates secure knowledge. But it is only for the high priests of science to exercise this magic. The common mortals are but bystanders to applaud (and pay for) the wonders of science.

When pressed, scientists are usually quite willing to admit that their methods are limited and hence their results are uncertain. However, when they are not pressed on it, for example in their popular science writings, these substantial caveats all too often end up being swept under the carpet. It is much easier to write about the fascinating conclusions science has reached than to dwell on the countless caveats and limitations. Nobody wants to read about this. People want to hear results. Moreover, there is the self-interest of the class of researchers. Scientists are agents in a social game for money and power. They want to advance their case. So, it is much better to write about the victories and leave out the complicated conditions of surrender that nature insists on in exchange for giving up its secrets.

The problem starts when scientists fall for their own propaganda and develop a specialist's myopia. They start to believe the myths they themselves have helped to create. Some, blessed with a magic pen, are not content to write about their own subject field alone, but start to preach on issues they know very little about. They extrapolate from their own field of expertise, whatever field of science it may be, to the world at large. So, they start to write about politics or God or religion.

Technological success paired with a healthy flow of good stories has helped to establish science as one of the mainstays of contemporary popular culture. Science at the centre of society is a good thing, as long as it is balanced by other philosophies. It seems to me that we are now entering a phase in the history of ideas where science slowly develops an overbearing weight on our societies. This has many reasons, but is eagerly helped and propagated by a vocal minority of scientists who form the new movement of SR.

SR is about to be incorporated into our *zeitgeist*. The telling symptom is that belief in the scientific method is pervasive in popular culture. Criminal justice bases its verdicts on science. Governments keep councils of Professors for advice. The future of western advanced knowledge economies is believed to depend on science.

One of the telltale signs of SR's encroachment on every aspect of life is that we are increasingly losing the ability to accept things that seemed quite natural to our not too distant ancestors. Non-scientific reasoning becomes unimaginable. How outrageous does it seem to us now to suggest that DNA evidence technologies may not be a good thing. Of course we want it, self-evidently. How shocking is the thought of being treated by a witchdoctor. How unthinkable it is *not* to organise the public health-service in accordance with scientific knowledge. How stupid does it seem to believe in a God with a white beard sitting in the sky somewhere? What a silly thought it is to think that there are ghost or spirits around us that influence our lives.

Not surprisingly, a by-product of the rise of SR is the new atheism. The professed belief in the non-existence of God is now a cultural phenomenon that has long left the confines of academia. At least in the Anglo-American world it has entered into the mainstream, although it probably hasn't become a majority movement yet. Still, it is big enough to start to leave its mark on popular culture. The magician and mentalist Derren Brown only thinly veils his anti-religious agenda. Another example is the "Invention of Lying" an unfunny Hollywood movie with a lame, but openly atheist plot. Its protagonist, Ricky Gervais, also enjoys some success as a stand-up comedian. Again, he likes to litter his (not very funny) act with atheist references. What makes both Brown and Gervais interesting is that they clearly have a missionary agenda. They do not only want to attract viewers and make money, it seems, but they also want to help viewers to get out of the prison of religion. They wish to enlighten them with the good news about science.

Non-belief in a God is one thing. Openly expressed and evangelical belief in the non-existence of God is different—and strange. There are many things people do not believe in, but still they do not feel the need to profess their non-belief publicly and constantly. The hysteria with which the atheist cause is propagated suggests that a culture-war has broken out. It is the epic struggle of good against evil. It is the battle of the force of rationalism and science against the dark armies of religion or God against immorality, hedonism and the armies of satan, depending on which side one stands on. Another way to look at the conflict is not to see it as a culture-war at all, but to see both atheism and fundamental religiousness as expressions of one and the same modern *weltanschauung*. Literal interpretations of scriptural texts by necessity come into conflict with science. However, these texts may never have been intended to be taken as factual statement. Fundamentalists who understand sacred texts as such just do so because their parameters of thinking are shaped by a scientific *weltanschauung*. Contemporary man has lost her ability to understand the meaning of symbolic texts and ritual and to think outside the dichotomy of truth and falsehood.

In her book "In Defence of God" Karen Armstrong makes the point that our view that religion is a set of particular beliefs in specific dogmas is an idiosyncrasy of our times. Most religion elsewhere was about mystery and symbolic understanding with little emphasis on set dogmas. Faith was an act of doing things, participation in rituals, not an exercise in professing manifestos. The development of the contemporary western understanding as instantiated in the new evangelical movement in Christianity and bible-motivated Creationism is a novel phenomenon that has the same philosophical roots as SR itself. So, when SR is a misunderstanding of science, Karen Armstrong makes the point that fundamentalist religion is a misunderstanding of religion. What appears to be a culture war between two opposed views of life, turns out to be a spat between brethren. Apparently for us contemporaries it is extremely difficult to think of myths and symbolic stories as both not literally true *and* meaningful.

Still, some may say, even if religious fundamentalism and science grew out of the same intellectual seed, there is still a battle to be fought. The two entities are brethren, but still enemies, just like Cain and Abel. Science, after all stands for free exchange of arguments. It is inherently free and egalitarian. Religion is not. It, so it is argued by atheists, encourages dogma and discourages independent thought. A society organised around the principles of scientific debate and scepticism would certainly be better than a Taliban-like theocracy. So, brethren or not, we should fight religion.

Then again, when an insight becomes too clear, a position too evident, then perhaps the time has come to switch on critical thinking. Let us have a closer look at this phenomenon of SR. One thing that strikes immediately is its quasi-religious tendencies. I have experienced over and over again that the zealous proponents of SR can be every bit as dogmatic as the religious fanatics they hate so much for their dogmatism. Working groups on evolution can be every bit as religious as any madrasah.

Dogmatism and intolerance are built into SR. Its central tenet could well be "Thou shalt not have another way of reasoning." Science is the way and the truth and no one comes to knowledge except through it. SR is the claim of exclusivity of science. The other side of the same coin is intolerance. Proponents of SR may be able to tolerate criticism, but they they cannot accept those who do not understand that there is no other rational way than that of science. I always wonder why SR is so intolerant. Perhaps it is the ongoing culture war with religion that led to this radicalisation. Perhaps, though, it is a consequence of the origins of western science. After all it has emerged from a Christian culture, a culture that also felt it had to convert the rest of the world with the sword.

The intolerance built into SR manifests itself in the writings of its proponents. Specifically in the new atheist literature dispassionate scientific analysis and argumentation is often abandoned in favour of hot-headed rants against religions, their institutions and privileges. It is striking that some of the brightest and most imaginative people abuse their own good names and turn out anti-religious SR-pamphlets that are not only poorly thought through, but also intolerant, anti-democratic and reactionary.

SR is not about science. Indeed, within science there is no need for such ideologies. Scientists know very well that they need to do science and not God. SR is an ideology for society at large. It is a proposition for the *res publica*. The idea is that society should be organised along rational principles. So far probably everybody agrees. We certainly do not want society to be organised along irrational principles. The problem is that SR equates "rational" with "scientific," hence society should be organised along scientific principles.

At first it may be difficult to see what could be possibly wrong with that. A state ruled according to scientific principles must be a good thing, surely. Politicians should heed the recommendations of scientists who, devoid of self-interest can advise based on the latest knowledge and insights. Indeed, this is happening already. Government, courts, medical care, public administration, economics, use science to varying degrees in order to evaluate themselves, plan ahead, make decisions and so on. In a society based on SR, there would no longer be "varying degrees," but all aspects of life would be based on science.

The idea that a state organised along scientific principles would be a good thing relies on several premises. The first one is that research is value free. It merely provides factual information. Decisions can be made based on this. This is not true, however. Science is not value free. The moment society applies a rationality that is based on science a certain type of ethical reasoning comes as an inextricable part of the package. This ethics of SR may not be the one we would elect to have as freely reasoning moral beings. Worse still, after some exposure to this type of reasoning we cross the Styx and we are no longer able to see this new ethics as what it is. A value choice. We would see it as the only way of thinking.

Another premise of SR is that science covers all spheres of life and that the scientific method is applicable to answer any kind of question: If it is not science, it is not knowledge, or in any case not any knowledge worth having. What makes this premise very interesting is that it is rarely explicitly formulated, but often delivered subliminally, in-between the lines of popular science books or TV programs.

Most people when they talk about science probably have Galileian science in mind with its high degree of certainty. In reality, the kind of science that is used in the context of large scale decision making in society is a very different kind of thing dealing with real world complex systems. Science is not good at complexity. If decisions that affect many people are to be made, then scientific advice must itself be scrutinised. Moreover, decisions in the public sphere will not only depend on hard facts, but also judgments and intentions, the willingness to take risks and visions for the future.

Scientific reasoning cannot replace these soft-aspects of decision making. Whether or not a society wishes to use nuclear power is not reducible to a scientific question. Similarly, how we feel about genetically modified food, whether we prefer a *laissez-faire* economic policy or a Scandinavian style social democracy is a value choice that cannot be settled by empirical data. Science can only be one part of societal decision making. SR would not do well as a universal principle to govern the state.

Another key assumption or tenet of SR is that science is a universal, near limitless, knowledge generation machine. Whatever question you have, whatever phenomenon there is, ask science and you get an explanation or even prediction. The tenet of SR is that it covers all there is in the world. There is nothing that cannot be analysed scientifically.

In reality, science is clearly powerful where it is applicable, but has very little to say about most other things. Away from the Galileian highway of science, science is not even a gravel road but made of poorly marked dirt tracks through some unknown wilderness. The vastness of the unknown territory is rarely talked about. How often do we see a scientist write about the things she has not found out, in fact the things she is not even planning to find out? Popular science communication always has an upbeat feel to it, prophesying the nearing salvation of mankind through science. It gives us reason for optimism that all will be well if we just believe in (and continue to pay for) science. Ignoring these messages of hope, if we look at how things actually are, we cannot help but be happy for what science has achieved so far. At the same time, when we evaluate with a sober mind what we do not know as of yet, then we have to be much less optimistic about.

Certainly there is promise. Science will continue to make many discoveries, probably even many unexpected and surprising discoveries. Nobody can predict what we will know in 100 years time. It may well be that some clever mind will have figured it all out and discovered how to do truly universal science. Maybe we will have solved the last problem. By the same token, it may well be the case that we will never get there, that the vast unknown wilderness will remain vast forever.

The assumption integral to SR is that open problems in science are just waiting to be solved. Eventually everything that seems mysterious today will fall out as a trivial corollary from a future theory. A case in point is complexity theory. It may well be, so the proponent of SR will argue, that at present we cannot deal with complex systems very well. However, this is because of technological and scientific limitations that will eventually be overcome. Such optimism is engrained in SR, but is rarely explicitly stated. Throw out the optimism and you need to abandon SR. Clearly, there is no rational justification for this trust in eternal progress. It is faith. Indeed, faith that is very much contradicted by empirical facts. While the volume of scientific output increases at an exponential rate, there has hardly been a quantum leap in our understanding of nature during the last 40 years or so. Technological prowess increases and the amount of detailed knowledge we produce grows, but there is no substantial progress that gives us any reason to hope that during the next 40 years there will be a fundamental break. There may well be one, but equally there may be none. We do not know and we have no evidence to go by.

One argument that is often made by proponents of SR is that applying science in all walks of life is rational, because the scientific process is rational. This is a somewhat subtle argument, so care must be taken when analysing it. First of all, it must be said that the premise is likely false. Science is not a rational enterprise. It is based on a number of assumptions that themselves cannot be rationally defended. To a large extent science is a social process and its methods are a mix of what has been shown to work reasonably well, what is technologically possible, what is needed and what is socially accepted.

Even if science was a perfectly rational activity, SR is not. There is a considerable amount of irrational faith underlying SR, reminiscent of religious cults. Just consider the system of myths that comes with SR. Stories that are recounted over and over again by the scientific community for themselves but crucially also for society at large. My favourite one is the Genesis myth, the epic struggle between Galilei and the church. It has developed into the creation myth of science itself. It is the story of a man who single handily pulled us out from the ignorant and dark pre-scientific age and founded the new eon of reason. There are countless versions of this strewn all over popular science literature (including this very book), re-iterating the deeds of the great and beloved founder of modern science. The tales are about the tenacity of the prophet of reason. He did not give up even when the enemy persecuted him and ordered him to recant.

As so often with myths, there is actually a kernel of truth to the story. However, what is often described as the courageous struggle of good versus evil, reason versus religious extremism, may have been quite different. A sober historical analysis show that the reality of the church proceedings against Galilei was a very different from what many imagine. Historians (Mayer, 2010) now understand Galilei's dispute with the church as a routine legal case. It was not so much a long march to freedom, but a juridicial spat. Not quite a trivial one, but then, not of major importance either. What is more, Galilei himself seemed to have conducted himself clumsily which gave him unnecessary trouble with the authorities. There is a marked dissonance between those who continue to propagate the creation myth of science and the cutting edge historical knowledge on the subject. Then again, why let the truth get in the way of a good story?

At first it may seem to be hard to understand why authors who defend SR and the importance of empirical research can ignore the state of the art of our historical understanding. SR is precisely the idea that research should not be ignored, including state of the art historical research. So, why then do proponents of SR continue to spin their own yarn when it comes to Galilei? One possible answer is that they simply are not up to date with this latest knowledge. This is possible, but would be embarrassing. The second possibility is that they do not care, because the story about Galilei is not primarily meant as factual information, but to say something about science. It is meant as a creation myth, or at least as a symbolic tale that reflects the struggle of humankind against the powers of religion.

Proponents of SR do not only tell one another creation myths they also developed a canon of stories that show the supremacy of science over other modes of reasoning. Followers reinforce their faith in the omnipotence of the scientific methods by telling tales of how science has overcome superstition and replaced it with the light of proper knowledge. The pattern of the stories is always the same:

Pre-scientific man did not understand phenomenon X so

ascribed it to some divine action. For example, lightening is the ire of Zeus.

Science then discovered that X is a specific instance of a larger class of phenomena that can be described theoretically. Lightening is now understood as a manifestation of electromagnetic phenomena.

Hence, science has once more replaced Gods with a rational explanation.

This interpretation of scientific development is frequently found in popular science literature. Indeed, there are entire books published that are just collections of stories of this kind. One may wonder why authors feel compelled to tell them. Probably they find them amusing. They certainly make us feel superior. Our ancestors were so ignorant and stupid to fall for myths that seem so ridiculous now. How nice that we have progressed beyond that.

There is no doubt that it is a good thing to understand the physics behind lightening or any other natural phenomena. At the same time, it would be a misunderstanding to assume that the ancients thought of the "ire of the Gods" as an explanation for lightening in the same way as we think of electrodynamics as an explanation.

Things are never that simple. For one, we must remember that the myths of pre-scientific man were not meant as scientific theories. They were not intended as precise accounts of factual events or explanations, but rather as parts of a larger symbolic universe in which ancestors lived. As a society we have largely lost the ability to appreciate non-factual symbolism, but not entirely. Particularly children's stories, such as Grimm's fairy tales are rich in moral and social contents. They are not just stories in their own right, but have social knowledge and norms embedded in them. Brand identification in adults is a remnant of non-factual truths in the modern world. Marketing departments in any bigger company try hard to associate brands with values.

This gives the consumer the opportunity to identify herself with those brands by buying the product. Commercials are the little myths that go with it. Nobody is foolish enough to think for a moment that the little stories told in the advertisements are anything but fiction. Nonetheless, they help to create meaning whether we like it or not.

The stories about old myths and how they are overcome by science are just that: stories. They compare apples with pears, scientific knowledge with other types of knowledge. Historical science is very much aware that the inner worlds of our forefathers were different from ours. Taking ancient myths as factual knowledge is simply a misunderstanding of ancient myths and their social role.

Some may feel that I overstate the case. Maybe, these stories about science and how it overcomes superstition are just harmless entertainment, told at cocktail parties by delegates who do not know what else to discuss. And, of course, this is a function of those stories, but they do transport an SR ideology. They are nearly always about some religious naivety, but rarely about the colossal mistakes that have happened within science itself.

Here is some circumstantial evidence: Nobody ever talks about the ancient Greeks. They have had phenomenally bad ideas (and some good ones). They are never ridiculed, because they are considered as the forefathers of science. Even in modern times there were many episodes where science has gotten it seriously wrong. My favourite example is that of Johann von Neumann. Apart from coming up with the concept of the cellular automaton, he was also an expert in quantum mechanics. Indeed, he wrote an influential textbook on the topic. In the textbook he proved a theorem. It was about an important but somewhat technical detail and he got it wrong. The proof was incorrect. That in itself is nothing noteworthy. Scientists, even the best ones, get things wrong very often. This can happen and is not a drama.

This case was different though. Johann von Neumann's error was elementary. Every graduate student of quantum mechanics should have spotted it. Still, it remained undiscovered for decades. Year on year, scientists took on faith what the great Johann von Neumann said, apparently nobody spotting the error until John Bell many years later finally set things right.

This is a case where science was simply believed. Where the community of researchers was persuaded by authority, rather than by the better argument as it should have been. As a story this is just as amusing as the stories about our forefathers who thought that lightening was an expression of the ire of the gods. Still, you never read about it in popular science books. These stories do not support the case of SR. They are amusing, but they are not useful to propagate the science myth.

You never hear about scientists erring, because the stories about myth debunking are are not told as entertainment, at least not only for that reason. These stories are weapons in a propaganda war. They are meant to discredit religion or any other perceived enemy of science as misguided. The von Neumann story does not discredit religion, nor communicate the good news. It does not bring the message of hope, optimism and progress that SR likes to tell.

The stories you hear are about science and how it has managed to explain all these previously mystical and mysterious phenomena and exposed them as corollaries of one or the other scientific theory. This should tell us that whatever myth remains, whatever it is we do not understand now, science will get there; if not tomorrow, then definitely very soon. It is meant to reassure us that there is still progress to be made, that there is nothing beyond the grasp of science, that the best is yet to come. It instills a belief in the everlasting power of science.

The stories never really say that science will keep discovering forever. Nowhere will you find the statement: "The best is yet to come." Still, by telling the stories of the liberation struggle of science promises are made implicitly. It is clear that this message of hope has to be made through stories. It is not possible to make a coherent argument that shows that "The best is yet to come" is a true statement. We cannot prove that science will eventually solve every problem. So, stories are needed to make the case.

SR cannot be defended rationally. As it turns out, there are other ways of promoting it as an ideology. Stories with subliminal messages. What an irony. The ideology whose tenet it is that everything should be accepted based on evidence alone, can itself not be established by rational argument.

Responsibility

SR is irrational and untenable. However, this does *not* imply that we should abandon science, stop funding research and become eremites or mystics. It does not mean either that some sort of opposite of SR is true. I am not arguing for irrationality, and I am not suggesting that we should stop using scientific insights in our everyday lives, including public lives. A scientific outlook on things can be good and is often useful. What we must avoid, as a society, is to fall into a pseudo-rationalist trap, where we narrowly define knowledge as scientific knowledge and where we consider scientific reasoning as the only valid type of reasoning.

SR is dogmatic, intrinsically intolerant and irrational. It is a blind belief in the universal power of science. It would not be a good thing if SR became a mainstream majority belief in our society. This does not mean, however, that science itself is dogmatic, intrinsically intolerant and irrational. Quite to the contrary. Scientific knowledge has a potentially equalising effect. Unlike class, nobility, sacred knowledge or money, it is accessible to everyone. Built into the system of scientific research is the requirement that evidence should be overt. Science opens itself up to critical analysis. When a research article is published, then the reader can demand to see the raw data that has been used to reach the conclusions. Translated into the public sphere this principle of openness restricts the ability of individual groups to build up a monopoly of power. Science helps to hold the powerful to account.

However, science also needs to be held to account itself. It must be clear when science can be used and when it cannot be used. Scientific advances should not always be welcomed as societal advances or progress. As users of science we must reason about the role of science in society, its ethical implications and societal effects. We do not want to be ruled by science, but want to use its benefits to our best advantage. This requires non-scientific reasoning about science and its uses.

The impact of science on society and its values can be quite dramatic. Take as an example the modern computer system. Computers themselves are of course only one of the products of science, but they are not neutral helpers that let us do all the things we have done before but more efficiently. Computers vastly expand the scope of things we can do, the amount of information we can process and the control we can exert. Yet, they come, like most applied science, with a whole package of implicit values and value choices. We are no longer free not to use computers, once we have them. They cannot be uninvented. Apart from giving us the internet and lots of games and more efficient ways to process text, computers are also efficient tools for the surveillance of citizen. Modern artificial intelligence and data mining algorithms create the transparent citizen. We are monitored and assessed by machines on an ongoing basis. Supermarkets can record sales down to the level of the individual check-out, business can monitor employee performances in great detail. Governments can control the economy and assemble large amounts of data about their citizens.

The impact of technology on life and values needs to be understood. It is not simply a question of science to decide what the next development will be, there also needs to be a way to assess what the impacts are and how they can be dealt with in the context of a democratic society. Questions of privacy and protection of citizens from undue surveillance are well known problems.

A radical program of SR implies blind faith in science and by extension technology. However, the enormous impact of technology on our lives and our societal relationships, together with the normative effect of inventions once made, requires a democratic control of science and technology. We have to decide what we want to invent and what types of technologies we prefer not to have. This decision is not a scientific question, but requires fundamentally different types of reasoning. If we live in a world ruled by SR, we would lose the ability to critically assess science and technology, because we would have denied ourselves the ability to reason non-scientifically.

Je n'ai pas besoin de cette hypothèse

SR and the new atheism are two movements that come together very often. Not every atheist is a proponent of SR, but nearly every proponent of SR is an atheist. Moreover, atheists, when challenged, often refer to ideas that come straight out of the SR cooking book. The connection is clear. SR insists that all there is is science which, they argue, is difficult to reconcile with the idea of a God.

There are at least two basic arguments that are commonly used. The first one is as follows:

1. People used Gods to explain natural phenomena.
2. One by one these natural phenomena are now explained by scientific theories.
3. Phenomena that have not yet been explained will be explained by future science.
4. Hence, there is no need to assume the existence of Gods to explain natural phenomena.

This argument does not strictly address the existence of Gods, but actually argues that nature does not point to Gods. This line of argument is directed mainly against the so-called teleological argument for the existence of God. This teleological argument is actually a whole family of arguments. The premise of the argument is that the world is not random, but ordered. Organisms, for example, are highly sophisticated machines. It is difficult to believe, so the teleological argument goes, that this order has not been brought about by some external agent. This is often illustrated with the watch-analogy. A watch is a highly structured and non-random object. Whenever we see one, we assume that somebody has made it. We immediately infer the existence of a watchmaker. The cogwheels and the mechanisms will not just fall into place by themselves. Organisms, even the simplest ones, are much more intricate than a watch. Hence, if we assume a watch needs a watch maker, then an organism needs an organism-maker, that is a creator God. There are countless variations of this argument, but they are all based on the same idea that order does not arise spontaneously.

Quite rightly, the proponents of SR point out that the watchmaker analogy is flawed. For one, we know now that spontaneous creation of order is fully compatible with the laws of physics and can indeed be frequently observed. No mystery there. Moreover, we now have theories that propose mechanisms for the generation of order over time. The theory of evolution is the most favoured one at present. The teleological argument is therefore to be rejected. Just because there is order, this does not mean there are Gods who created the order.

It should be pointed out that the rejection of the teleological arguments does not depend on the theory of evolution being correct or valid. Even if we had no theory at all to explain order, the existence of a watch does not imply the existence of a watchmaker. It only implies some (potentially unknown) mechanism by which the watch came into existence.

The teleological argument is not only flawed logically, but also scientifically empty. The problem is that, nearly by definition, the Gods themselves are unexplained. It therefore does not explain anything to point to Gods as the ultimate cause of organism. Saying that there is a divine watchmaker just replaces one mystery (how did the watch come about) by an arguably bigger one (how did the watchmaker come about). This latter point is the crucial death-blow to teleological arguments of any kind.

A modern version of the teleological argument is an allegedly scientific theory called "Intelligent Design" or ID. ID is an attempt by the Christian right to present a more palatable version of outright creationism. To be fair, in ID-writings one does not frequently find the word "God," but equally the reader is not left in much doubt about the intentions of the authors. ID is a whole series of arguments whose structures are quite similar one to the other. First, some biological phenomenon is presented. Usually a mechanism that is central to the function of a living being. Then, it is shown that the mechanisms cannot possibly function without each of its components being in place. Michael Behe, one of the main proponents of ID, calls this irreducible complexity. This insight that irreducible complexity exists is then used to argue that it could not possibly have arisen through evolutionary processes. Proponents of ID point out that evolution depends on a number of small changes whereby each step leads to a small improvement of the system. Irreducibly complex systems, so they argue, do not work unless all components are in place and fully functioning. Therefore, no sequence of small changes can give rise to them. Based on this the proponents of ID conclude that evolutionary processes are not sufficient to explain the existence of irreducible complex systems. Hence, an intelligent designer must exist.

The reader will see the flaw herself now. Even if we accepted that a mechanism cannot be explained by evolution, this does not mean that a designer exists. It could well be that there is another natural, albeit currently unknown, explanation for the mechanism in question. Moreover, it is not correct to say that irreducible complex systems cannot arise by natural evolution. To see this consider as an example one of those old bridges made of bricks. When you walk under such a bridge, you will see an arch, a so-called brick arch. The special feature of this construction is that the individual bricks in the arch are not supported by bricks underneath them. Instead, they are held in place by all the other bricks in the arch. The weight of the bridge overhead presses the bricks together and thus prevents them from falling out. The construction can only work when the arch is complete. This is very much like Behe's mechanisms of irreducible complexity. Take out one brick and you have no arch. So, the question is now, how can such an arch be built. Unless you put in all bricks simultaneously, which is clearly not possible, you cannot have an arch. Still arches exist. Have they all been created by an intelligent designer?

Well, yes they have, in the sense that all arches are human-made. However, even the intelligent designer had to use a trick, a trick that is equally available to evolution. The builder of the arch had to construct a scaffold or some other temporary structure that holds the bricks in place until the last brick is set into the arch. Once this is the case, she can remove the temporary structure and the arch will hold by itself. Problem solved.

Evolution can use the same trick. A biological system that appears to be irreducibly complex system, may not have been irreducibly complex in the past. There could have been some other structures in place that allowed the system to evolve gradually, while retaining function at each point. Whether or not such scaffolding actually has occurred in evolution, is a separate question. The main point is that irreducible complexity by itself is not an argument against the evolutionary origin of a system. At this point, it is fair to conclude that ID is a pseudo-scientific idea that is not worth any further consideration. Similarly, all other teleological arguments are weak and stricken by similar problems.

The proponents of SR clearly do have a case when they criticise teleological arguments and consorts. These are bad confusions of the mind. Yet, SR does not leave it at that. They need to take the next step. Having pointed out that there is a lack of evidence for Gods in nature, they then go on to conclude from this that assuming the existence of Gods is implausible, and certainly unnecessary, irrational and stupid.

Let us go through the argument step by step. Having debunked teleological arguments they go on:

1 There is no direct evidence for the existence of Gods in nature.
2 Hence, scientifically it is not a good strategy to assume the existence of Gods.

So far, there is nothing to be said against the conclusion. I agree fully with their line of argument. Indeed, there is no scientific evidence for the existence of Gods. Moreover, given the nature of scientific research, assuming Gods is bad practice. Super-natural beings are not parsimonious ideas. At best they would be a whole new set of problems to explain and research. New questions are normally good news in science, because they lead to excitement. In this particular case, however, it would be the end of the road because Gods are chronically difficult to experiment with in controlled ways. Supernatural beings are simply not the kinds of things that make useful objects of scientific theories and are therefore best left out from them.

If the proponents of SR stopped here, it would be hard to disagree with them. Unfortunately, they do not. Instead, they take the next step and conclude the non-existence of Gods. The line of reasoning is quite straightforward, although rarely articulated explicitly. Starting with the idea of the universality of science proponents of SR typically argue as follows:

1 Science is the only legitimate source of knowledge. The only valid knowledge is scientific knowledge.
2 Hence, if no scientific case can be made for the existence of Gods, then Gods do not exist.

To be fair, the clever ones among the proponents do not shut the door completely. They admit that they cannot be completely certain that Gods do not exist, but hasten to add that assuming them is exceedingly implausible. They would be wasting their time pursuing the idea.

Be it as it may, the validity of the claim about the non-existence of Gods crucially hinges on the correctness of the premise of the universality of science. The problem is that the validity of the universality claim is not itself an empirical question, but a philosophical one. At least as far as our understanding of science is concerned, there is no reason whatsoever, to believe in the premise. Science is a very heterogenous activity, that works well in some domains and less well in others. It seems to be very good in the most fundamental sciences, but not so good as a way of understanding complex systems. Indeed, our ability to understand and manipulate complex systems is miserable.

What is more science is not a well defined process. There is no scientific method as such. Science is a social activity. It is organised in a way that ensures mutual scrutiny of results and arguments between members of the community. This is a way that has been shown to be successful in generating reliable and robust knowledge, although nobody quite knows why it works and certainly nobody can give a rational argument that it has to work. Science is a great thing, but it is far from generating eternal truths. What is more, there is no such thing as a unified scientific method. What counts as a good argument in biosciences is very different from what mathematicians consider a good argument.

Scientists are trying whatever they can. In physics this works out to be the Galileian paradigm, in biology it is the Monodian approach or stamp collecting, and in climate change science it is dealing with untestable models. When SR claims the universality of science, then one has to ask: What science? What is science? As long as this question remains unanswered, it is fairly meaningless to claim universality for the thing that we do not even understand.

The probability of God

Some years ago there was a bus advert all over the UK saying:

There's probably no God. Now stop worrying and enjoy your life.

The advert is remarkable in at least two ways. Firstly, one may wonder about the motivation of people trying to convince others that something does not exist. It is opaque to me why anyone would want to spend money to simply to state that X does not exist. Maybe this is just me.

The second remarkable thing about the advert is the word "probably." It is clear from the advert that the authors are very confident about the non-existence of Gods, but at the same time, they implicitly admit that they might be wrong about it. The word "probably" is meant to express the residual doubt that is associated with any scientific statement. So, it suggests that they consider the existence of God as a scientific hypothesis that, after consideration, they felt compelled to reject based on its lack of plausibility. At least, even if Gods existed, they are so elusive in the universe, so irrelevant that it makes sense to ignore their existence.

What intrigues me about the whole discussion about God and science is that those who wish to "prove" the existence of God also like to use science. With doubtful outcomes. One common argument that is frequently used is the argument from fine-tuning. The idea is simple: We know that the small change in some key physical constants determining the strength of the fundamental forces such as gravitation or the forces between fundamental particles would result in a completely inhabitable universe. Make them just a little bit stronger and the world collapses, make them just a bit weaker and everything flies apart. Either way, there will be no Sun or Earth, let alone life. If the universe as we know it is only possible within a very narrow range of the constants, then it is extremely unlikely that just the right constants came about by mere chance. There must surely be somebody who has tuned the constants to make the universe and life possible...and we are immediately talking again about a watchmaker. We know already that the argument is invalid. An apparently fine-tuned universe does not suggest, let alone imply, a creator. Still, scientists are sometimes a bit uneasy about this fine-tuning and feel that there is a case to be answered.

One argument they came up with is the anthropic principle. It roughly says that, if the constants had not been finely tuned, we would not be here to wonder about why they are finely tuned. Even if the fine tuning we observe is extremely unlikely, so is any other configuration, but we would not have observed any of those. Sometimes this is complemented by theories that our universe is not just a one-shot game, but universes come into existence and disappear again, each time with different constants. These ideas lend themselves to superimpose an evolutionary process, whereby only the good universes survive for long enough, but the bad ones die again very quickly. This is mostly speculation, of course, and as far away from science as Genesis.

A different approach physicists take is to come up with theories that show that the constants have to be the way they are. While this certainly makes interesting physics, it does not mean anything. Any theory ultimately needs to be based on some contingent facts. So whatever its results, they have to be contingent as well. It is impossible to prove that the world has to be the way it is. There are many different kinds of universes that are imaginable. Indeed, the most plausible world is the one that contains nothing, no universe. An empty world.

A marginally cleverer way to use the fine-tuning as an argument for God is to drop the insistence on the implication. One may concede that fine-tuning does not logically imply the existence of Gods, but still somehow seems to make it more likely. One may reason that, if the world was created by one or several divine creators, they would certainly want to create a world that is interesting and therefore fine-tune the parameters to achieve this. On the other hand, if the values of the physical constants were not fine-tuned for a particular purpose, then they would most likely not lead to an interesting universe. There is no certainty in this conclusion. Still, at least according to some, our life-friendly universe points to a creator in the sense that it makes her existence more likely.

Probability arguments of this sort have been defended by a number of researchers, including the Oxford philosopher Richard Swinburne. There are two problems with this line of reasoning: Firstly, the case that X makes the existence of Gods more or less plausible usually employs a considerable amount of hand-waving. Secondly, and this is the more serious objection, it does not make sense to discuss the probability of the existence of Gods because we do not know the entire probability space. There is probably an infinite number of possible concepts of God. There is also an infinite number of possible laws of nature and most of the space of possible concepts and laws are unknown, both conceptually and in their consequences. It therefore makes no sense to speculate how plausible or likely it is that a universe with properties X_1, X_2,.. is created by Gods rather than the result of laws of nature. We simply do not know how many possible sets of laws there are and what kinds of worlds would emerge from them. Neither do we know the entire set of Gods and their intentions. It may well be that most of the Gods we could imagine are not at all interested in creating worlds. All we have is a single universe and a single understanding of the physical laws. What Swinburne is doing is like inferring the number of sides of a dice when he only knows that the result of one throw was 42. There is no way of knowing what the distribution is that led to this single random number.

We conclude, the teleological argument does not hold. Neither does a probabilistic version of it. We cannot derive the existence of God from observations of natural phenomena. Neither can we derive it from the laws of physics, biology or any other of the sciences. These laws do not even make the existence of Gods probable or improbable. They simply say nothing about this topic. In fact, the process by which the laws are created explicitly excludes all considerations of the supernatural or divine.

This means that science has nothing to say about the existence or indeed the non-existence of Gods. Whether or not there is a God, is not answerable by science. Gods are not useful scientific hypotheses and are therefore by necessity excluded from science.

A super-Tierra

The ancient Greeks believed that their Gods mostly resided on Mount Olympus in Greece. This is admittedly the highest mountain of Greece, but with just under 3000m to the summit it is still not that high. I always wondered why the Greek did not simply go up Mt Olympus to check whether the Gods are actually there. One could think that it was too difficult for the ancients to do this. This explanation does not convince me. There would have been no need to climb all the way to the summit. The Gods would certainly not all live on the crammed space of the highest point, but probably have their dwellings all over the higher regions of Olympus. Even for the ancient Greek it would have been relatively straightforward to send up an expedition to check for the presence of the Gods. They don't seem to have done this in ancient times, even though this would have been a nice empirical test for their religious beliefs. Apparently, the idea of experimentally testing religious ideas is a modern one that did not cross the minds of the pre-Galileian ancient Greeks.

Modern faiths are not quite as easy to test. The Christian God, for example, is thought to reside in heaven. This is a much less specific place than Mount Olympus. When air and space travel in the actual heaven began nobody seemed to find any God. Yuri Gagarin allegedly said "I see no God up here" when he was the first man in orbit. As a consequence the religious "heaven" become more of a metaphor, a placeholder for somewhere inaccessible. Nowadays, only a few people actually believe that God is sitting on a cloud in the sky.

Conceptually, it does not make much sense anyway to imagine Gods sitting in the physical heaven just above the atmosphere, or even on a mountain on Earth. If we try to conceptualise Gods as creator Gods that have brought the universe into existence, then one has to think of those Gods as being located outside our universe, somehow. At least there has to be a place outside our universe and the Gods have to be able to get there, otherwise they would not have been able to create the universe. It is difficult for us to think about things that are outside the universe, which is a realm of reality about which we cannot possibly know anything. It is therefore quite tempting to abandon speculating about this, as an ultimately unanswerable question.

There are some ways to think about it though, that could make sense. One possibility is to see the universe that we experience as an extremely sophisticated computer simulation. Imagine that we are all just parts of a final year software project of a computer science student somewhere in an inaccessible higher order world. Let us call her, the student, Maxime C. In fact, we do not know whether she is a man or woman or indeed whether these kinds of categories even apply in her world. The idea is that Maxime created a simulation a bit like Ray's Tierra but much more intricate. She created rules that are the laws of nature of her universe. Then she wrote a computer program that implemented these laws. We can imagine that Maxime did very much the same as Tom Ray, only that her simulation is much more sophisticated. Also, she did not start with self-reproducing entities, but with an initial configuration where all the simulated matter is concentrated in one spot. When she started the program, the big-bang happened. She then leaned back and watched her artificial world, our real world, evolve. As it happened, she chose the parameters, the numbers that we call the physical constants, just right, so that her initial bunch of elementary particles not only exploded, but evolved into a rich world. A rich world that amongst its millions of stars and galaxies also had one (or maybe more) planets that supported life, life that eventually started to get intelligent and developed the ability to wonder about its creator. If you buy into this scenario, then our universe is not an independent reality, but we are only a computer simulation, a science experiment. Our universe is then, at best, a second order universe, the creation of another reality. By the way, it may well be the case that Maxime is herself just a part of a simulation model and she may not even know. And of course, those who simulate Maxime may equally just be simulation models. Who knows where this ends, but we do not need to worry about that now.

One thing is certain. I do not feel at all like a computer simulated being. I am not made of zeros and ones, but of flesh and blood. I do not hear the fan that cools the super-computer in which we are supposedly all residing. So, the idea that we are but a simulation appears to be far-fetched. Absurd as it seems at first though, it is quite compatible with SR. Let me explain: One of the basic tenets of SR is that all natural phenomena are ultimately explainable by scientific insights or theories, either present or future. This basic assumption implies what philosophers call *reductionism*. There are many versions of reductionism and it would go well beyond the scope of this text to discuss them all. The basic idea of reductionism is easy to explain though. It is the idea that all phenomena in the world ultimately are reducible to some fundamental laws of nature. One usually thinks of these laws as the laws of physics. Chemistry is then just physics in the sense that every chemical reaction can be explained by some basic physical theories. Similarly, biology is just chemistry (albeit a complicated type) and psychology is just biology and social sciences ultimately should be reduced to psychology. In reality there are a number of well thought out versions of reductionism, but they all share this basic idea.

The laws of physics are always formulated in mathematical terms. If we believe in reductionism, then this means that all the properties of our world can be ultimately expressed formally that is by means of computer simulations or mathematical formulas. This also means that they can be simulated on a computer. So, in principle, we could make a copy of our own universe. All that is required is a specification of all fundamental laws of nature plus a complete description of the world at time $t = 0$. Within our universe it would be very difficult, most likely impossible to implement a complete simulation of our own universe. However, we assume that Maxime lives in a higher order universe, so there is nothing preventing her from simulating us, or rather creating us in simulation.

One may object that the world does not feel as though it is but a simulation. We tend to think of the information processing that goes on in our computer as evanescent interactions of electrons. Computational worlds are virtual and not real. The Tierra organisms are just simulations of organisms, not made of flesh and blood like NiuNiu my cat. Then, again, we do not know what it feels like to be a virtual organism only. Or, maybe we do, but do not know that we do. Maybe we just do not know what it feels like to be non-virtual.

The fundamental idea of SR is, must be, that physics or at least some future physics, gives an accurate representation of reality. SR has to assume that all there is to know can be represented formally. Consequently, it must be possible to calculate the fundamental laws of physics and hence to implement a virtual reality that behaves just like our own reality. We can probably not simulate our own universe in all its detail, but we can simulate simpler, still lower order universes. I am not certain that our simulation models will ever support sentient beings, but this does not mean that assumed simulations in assumed higher order worlds could not. If we sentient inhabitants of our universe—including our alleged intelligence—are just the consequence of the evolution of a dynamical system defined by the laws of physics and an initial condition, then there is no reason at all to reject the idea that we could be just simulated as well. Flesh and blood or bits and bytes would be equivalent.

When sentient beings emerge in a computer model, the only thing they could experience would be the informational world that they inhabit. This reality would have to seem real to them, just as our world seems real to us. It is the only reality they know. Indeed, maybe we are them. We would not know, and have no way of knowing. These beings would be able to perceive and think about themselves, but they would not have access to the world outside their simulation, the higher order universe in which their own cosmos is simulated. We can only hope that Maxime will not abort her simulation anytime soon.

We can never meet the great simulator and find out about her, that is unless she somehow chooses to reveal herself to us which she may or may not do. For all intents and purposes the great simulator would act as a God for us. She created us in the sense that she created an artificial universe that supported our evolution. Perhaps she even created us in her own image in that we are similar to her own species in some ways. Maybe she is even amused about the fact that her own creations start to speculate about her. If indeed she exists, her world and her reality are also completely inaccessible to us.

Admittedly, this scenario is a complete fabrication. There is not a shred of evidence that we are a simulation. However, what makes the idea of Maxime so intriguing is that she would for all intents and purposes act like a creator God. She would be beyond science, unprovable, indeed incomprehensible, yet still real. In fact, she would be more real than we are, in a very definite sense. She would be invisible, undetectable by scientific means, she would be a creator, and Yuri Gagarin had no chance of ever seeing her. Yet, still, she is fully compatible with SR.

Limits

One of the basic tenets of SR is that if it is not scientific, it is not worth knowing, probably even meaningless. As a philosophical position, SR is untenable and it is in many ways a warmed up and dumbed-down version of the positivism of the Vienna circle, a philosophy of science from the early 20th century. In a nutshell the idea of this philosophy was that claims and statements are only meaningful when they are empirical. So, any meaningful scientific statement should ultimately boil down to so-called *protocol sentences*, that is sentences that directly express empirical observations. An example of such a protocol sentence is "Dr. X has observed that the temperature of the solution was x degrees." While positivism initially became an influential philosophy of science, it turned out fairly quickly that its basic tenets are not defensible. Scientific statements can usually not be reduced to protocol sentences. For example, generalisations (i.e. "falling objects follow the square law") can never be derived from a finite number of observations. This means that science needs more than just protocol sentences. Moreover, positivism itself, that is the requirement that all science is empirical is itself not an empirical theory.

Admittedly, proponents of SR probably no longer believe in

all the tenets of positivism. Still, the spirit of their ideology is not so dissimilar. Proponents of SR always insist on testability and they do so in all walks of life and without compromise. This seems to be an inbuilt reflex of scientists. Philosophers of science know fully well that positivism and hence SR are antiquated. Just as positivism is not an empirical theory, SR is not a scientific theory. Its own truth is not testable. This makes SR a question of belief, so more akin to a religion than a scientific theory. What an embarrassment!

One may wonder why such an obviously outdated philosophy of science can possibly gain such traction when philosophers of science know fully well that there are lethal arguments against SR. The answer, I think, ultimately comes from the fact that science is a social phenomenon and scientists form a class within society at large. As a class they need their own propaganda tools both to convince the outside but also to create coherence internally. Hence the myths, hence the stories. Only, scientists sometimes get persuaded by their own spin and start to take the myth as truth.

Another reason is that scientists sometimes feel drawn to philosophy. Many of those, no matter how good they were in their own fields, turn out to be rather poor, ill-informed and amateurish philosophers of science. Still, they may wield enormous respect within some communities. Their words may even percolate into society at large through their popular science texts. It is not that they are not very intelligent people, but they are usually not trained in philosophy. Still, intelligent people can say stupid things. There is no reason to assume that an evolutionary biologist, no matter how famous she may be, would also make a good philosopher of science. Strangely enough though, scientists who are excellent in a particular field often seem to feel a calling to exude their wisdom in other fields as well. The results are often rather embarrassing. The assumption that Prof. X is an expert in philosophy of science because she is an expert in some other field is a very strange phenomenon. Would you think that a physicist, say a Nobel laureate, would also be a good neurosurgeon? Probably not.

In this chapter I have tried to make the argument that SR is unsatisfactory because it is self-contradictory. The upshot is that SR only accepts scientific arguments, but is itself not a scientifically testable theory. This argument is somewhat theoretical and maybe a bit subtle as well. In the remainder of this book I will therefore make a different case. I will show that there is knowledge that is both meaningful and *in principle* beyond the reach of science.

I need to stress here the importance of the qualifier "in principle." There are many things that are currently beyond the reach of science and seem likely to remain so forever. For example, we cannot control the weather. It would be nice to be able to press a few buttons so that the next day it is 20 degrees with a light shower in morning. Or, we do not know how to halt ageing. Some may think that it would be desirable to give some drugs to people so that their natural ageing process is halted and they remain within the body of a 20 year old forever.

It is unlikely that we will ever have the technologies to do these things. Indeed, it is unclear whether we should have them. In any case, I cannot be certain. It may well be that in 100 years we can halt ageing or control the weather or both. I doubt it, but then again, I would not be the first one whose predictions have proven to be completely wrong.

This difficulty of predicting the future is often exploited by the proponents of SR to throw insult and scorn at those who had previously predicted the impossibility of knowing X or manipulating Y at those who declared that a phenomenon Z is a direct manifestation of the Gods, only for science to prove them wrong. X is now known, Y manipulated and Z explained. If someone predicted a new limit to science, then she could immediately be accused of invoking again the God of the gaps.

To be clear, I am not interested here in making prophecies. I do not wish to predict what will and will not be possible in the future. I cannot do this. My aim in the remainder of the book is a different one. Rather than merely stating that there is a problem that is so hard that science will never solve it, I will present a problem that is not only hard, but most of all impossible to solve on principle grounds. It is not impossible for science to solve because we lack the tools or because it is very hard to understand. It is impossible to solve because it requires arguments that are inadmissible in science. I am interested in finding a concrete example of knowledge that is in principle beyond the reach of science. Something that, independent of any progress, will never be understood scientifically.

One example is the existence of Gods. It is built into science that Gods or any type of divine or super-natural being cannot be incorporated as scientific knowledge. Even if there existed a God, she would never make a meaningful object of scientific study. A scientist has to assume the absence of Gods. So, Gods seem like a good case study. The problem with this, however, is that we cannot be certain that Gods indeed exist, which makes it a bad, indeed very bad, example. Another related example is the super-Brown, the magician who is not only an illusionist, but a real magician. Scientifically, we would never be able to prove that she really is a magician, but we would always have to assume that it has so far not been possible to understand her tricks. Again, super-Brown does not exist, so this example is not that good.

However, there is some other knowledge that every one of us has. Knowledge that is certain and cannot be consistently doubted. At the same time, this knowledge is of a sort that *in principle* is beyond scientific scrutiny.

Chapter 7

Je pense donc je suis

Unbewußt höchste Lust
Tristan und Isolde, Act 3

Jake and Captain Kirk

Towards the end of the movie Avatar, when the forces of evil have been beaten, the makers of the film decided to finish with an emotional feel-good climax. The scene we were allowed to witness was the transfer of the protagonist Jake from his human body into that of one of the blue giant people. To be precise, it was the transfer of his soul, or spirit or personality or whatever you want to call it. The human was allowed to become a full-blown honest member of the Na'vi tribe.

Avatar is littered with subliminal messages of the political and religious sort. This includes a strong critique of science and technology. At the end of the picture the protagonists even opt out of their high-tech society that has lost its moral grounding to live on a tree with blue people. Within the polite circles of SR, this kind of techno-criticism is very non-PC. However, what gets the tempers of the proponents of SR really boiling over is the idea that you could transfer people between bodies.

There is an implicit dualist philosophical position underlying the body switch scene, namely that the body and some fundamental essence of a person (FUDESS) are two distinct and separate things. The specific form of the transfer ritual of the Na'vi suggests that the FUDESS is a non-material thing. They did not perform any form of surgery, they did not cut out a part of Jake to implant it into the blue person. In the movie it seemed more like a ritual that led to a transfer of some sort of spirit, the soul if you like. So, following Avatar, the FUDESS seems to be non-material. At least it is something that can go from person to blue person, presumably back as well, without requiring any physical intervention into the bodies between which the FUDESS is to be transferred. According to Avatar, bodies are mere machines that are controlled by an apparently non-material FUDESS.

In the cult-series Star Trek there is a transporter that can take people from a special room on board the USS Enterprise and transport them to a different place, for example to another ship or to the surface of a planet. With this being Star Trek the device comes with its own theory. The creators of the series seem to imagine it as a combined scanner, information transmitter and remote 3-D printer. First, the atomic arrangement that makes up the object to be transported is scanned. Then the person (if it is a person; the transporters can also deal with non-living matter) is disintegrated and the scanned information is transmitted to the target location where the object is re-assembled atom for atom. It is not clear how precisely the reassembly process works, but we need not worry about that here. The main point is that, somehow, the transporter is able to make an exact replica of the original. The basic working principle of a transporter is a bit similar to what some very rich Americans sometimes do. When they come to Europe and take a fancy for one of the castles then they simply buy it, have it disassembled, ship it to the new world and then have it rebuilt brick for brick on their ranch in Texas. A transporter does a similar thing, but for Captain Kirk and his crew.

Given this operational principle of the transporter, it seems unlikely that the machine is able to transport a non-material FUDESS as well. All it does is gather information about how matter is arranged, and then it reconstitutes this matter again. Still, Captain Kirk and his astro-platoons when beamed to a planet and back appear to preserve their personality and memory and individuality after using the transporter. They seem unchanged and certainly unharmed by the teleportation process, even though technically they went out of existence for a short time between the disassembly and reassembly stage of the transportation process.

We cannot be sure how the fictitious transporter works, but from what we know about Star Trek it seems reasonable to assume that the teleportation device does not actually transmit atoms, but rather information. So when Kirk is transported to the planet, the atoms that made up his body are not physically moved to the planet. Instead the machine probably just scans the atomic configuration of Kirk, then dissolves him and transfers the information. On the planet Kirk is reconstituted with different atoms using the information that was generated originally. We cannot be absolutely certain that the Star-Trek beamer functions according to this principle, but at least existing teleportation technologies work in this way. So-called quantum teleportation experiments make exact copies of particles by simply transferring their exact state, not the actual material component. This is not science fiction, but hard science. Unfortunately, so far it only works for individual particles.

In physics an atom of X is the same as any other atom of X. Individual particles of a given type behave in exactly the same way if they are subject to exactly the same environment. They have no identity. Atoms of a particular type are not marked in any way, they have no names, no history, they only have properties. There is no Simran the carbon atom or Raj the Helium atom. There are only Carbon atoms and Helium atoms. Hence, there is no need to transport the original Captain Kirk atoms. They are not necessary, and they do no good when teleporting Captain Kirk from A to B.

Avatar and Star Trek have apparently quite different implied theories of FUDESS. If teleporting Captain Kirk really preserves his inner essence, then the FUDESS must be something material, somehow encoded in the atomic configurations that make Kirk. Indeed, it would have to be something that can be fully encoded as data. Otherwise, transporting would not work without doing damage. This again implies that there is no spirit, no soul that can outlive the body. The essential me or you, or at least what I and you believe it is, would only be a result of matter and its interactions. What we believe to be our inner life, our FUDESS would only be the flickering of neurons in the brain. Jake in Avatar, on the other hand, seemed to have transferred an immaterial self, a true soul without making any apparent changes to the physical body itself. The underlying assumption, therefore, must be that the FUDESS is something immaterial, some spirit or soul.

These two views of FUDESS are substantially distinct. If we follow a non-material view, as in Avatar, then this has some interesting implications. You could, for example, switch body with anybody, say with your neighbour. If she agreed to the body swap, you could then be able to control the movements of her body, see through her eyes, walk on the street and be recognised as the neighbour rather than yourself. If, during the time you are in her body, somebody stuck a needle into your neighbours arm, you would feel the pain. Indeed, you would feel all the sensations that your neighbour would feel normally. Similarly, your neighbour would be in your body and experience everything, you feel. It would therefore be a good idea to time such a swap so that it coincides with a trip to the dentist, although this may sour relations with the neighbour.

So, before continuing, let me now clarify what I mean by FUDESS: We all have a unique first person experience[4]. We experience the world, we can sense things, we can react, we can think, we can perceive pain, we can feel pleasure. When we die then the lights go out and we cease to exist. I define FUDESS as this thing that feels pleasure and pain, the thing that perceives, that thinks, the thing whose lights go out when it dies, the kernel of the subjective experience. The FUDESS is only observable by itself. You, dear reader, can infer my FUDESS, but you cannot observe it directly. You can see my actions, read a record of my thoughts, you could also speak to me, but you cannot directly experience my FUDESS. Even if we swapped bodies (which is probably not possible), you would still not be able to share directly my first person experiences. In this case it is then you or rather your FUDESS who experiences being in my body, you who sees colour through my eyes, or feels the dentist drill a tooth and so on.

Confusions

Body switches are a popular topic for Hollywood films. Mother swaps body with daughter, wife with husband, beautiful cheerleader swaps with high-school geek. The appeal of such films comes from the idea that people can explore the world, literally, through different eyes, which offers countless possibilities for humour. They usually culminate in some sort of catharsis, in that the characters that have swapped bodies see the world from a different perspective, which gives them a new, better attitude. Only when this is achieved are they allowed to swap back into the original body. What is interesting with Hollywood FUDESS-swapees is that they take their memory, including their auto-biographical memory with them as they move between bodies. Jake in Avatar still remembers his life before the swap. Similarly, the cheerleader who knows everything about herself pre-swap, but nothing about the scorned geek whose body she now lives in. Indeed, the movies would not work without this. Imagine you swap body with your neighbour, but once you swap you have lost your auto-biographical memory. Imagine, you would only have your neighbour's memories. If this happens then you would not know that you have lost your own memories. You may remember having agreed to switch bodies (since this is part of your neighbour's memories), but since you have no access to the memory of your former self you cannot be certain that the switch has actually worked. This means you should feel as though you have always been your neighbour, rather than yourself. Your neighbour would feel quite similar. It would make a fairly pointless movie. The cheerleader would think she has always been a geek and the geek would think he has always been a cheerleader. Life is as it always has been.

I would therefore expect that not much of your personality would survive a memory-less swap of this kind. Certainly, any difference of behaviours before and after the swap would be quite subtle, probably undetectable. Under the influences of the geek's memories, and his particular brain chemistry, maybe the cheerleader girl turns into a geek herself, straight away. Much of the character of a person depends on her past and hence her memories. If you have once been bitten by a dog, you are probably afraid of them. If you had very good experiences with dogs in the past, you are more likely to be drawn to them. Behaviour is also strongly influenced by physical effects, the balance of hormones in your bodies and your nutrition and health. Your neighbour's body will have a different cocktail of hormones and other chemicals which will modify your behaviour relative to your previous behaviour in your old body.

Without going into further detail, it seems that a FUDESS swap that does not swap memories as well is difficult to recognise as a swap. The swappees would probably not know that they have been swapped and they would not preserve their character into the new body, certainly not all of it, probably none at all. In any case, they would make very confusing movies. Film-makers understand this point very clearly and make sure their protagonists retain all their memories from the past.

Such body-swap gedankenexperiments reveal conceptual problems of a non-material FUDESS. Quite clearly, behaviour, memory and character all have an undeniable physical grounding. For example, memory can be impaired by various neuro-degenerative diseases. What degenerates is a part of the brain, but with it go memory and personality. Diseases or accidents can lead to amnesia. Mental diseases such as depression or schizophrenia can at least partially be treated using drugs that re-balance the brain chemistry. Brain damage can lead to severe distortions of the personality with often bizarre effects. Latest insights in neuroscience seem to suggest that even ethical choices are largely controlled by particular areas in the brain. Pin-pointed damage to the ethics area makes it impossible for the patient to make certain types of ethical choices. It is hard to avoid the conclusion that most of the things that make you *you* are directly caused by material interactions, leaving little for any mysterious non-material essences to do. Any Hollywood-style body swap therefore has to be thought of as memory-less, because the things somebody remembers are stored within the brain which is not swapped.

If the personality, the essence of a person was largely due to some immaterial substance, then it is hard to understand how it can be impaired by physical interactions, brain damages and the like. At present, we clearly do not understand how our internal worlds are generated and where they come from. Still, it is undoubtedly true that the brain and its physical state are a *sine qua non* for our mental lives and our personalities. No brain, no personality. Damage the brain, and you damage the mental capacities of the affected person.

If you are not convinced, just imagine that you swap bodies not with your neighbour but her cat or hamster or one of the spiders in her back-garden. Would the spider still behave like a spider when it is in your body? Would you, in the spider's body, still be human? It is difficult to see what part of your humanity can be meaningfully preserved in the primitive nervous system of a spider. It does not have the processing power to even formulate a single thought. Within this new body you would lack the computational abilities to realise your FUDESS, to process what you hear, see, and think. You could not be you. Now, take it even a step further. What about a free floating FUDESS, separated from the body? By definition, there would be no processing power at all. Hence, no thought, no feelings, no you. We conclude: A non-material notion of FUDESS does not seem to make any sense. The body-swap flicks are implausible.

The default conclusion seems to be that a Star Trek concept of FUDESS is more plausible. It certainly does not suffer from those particular problems. If we assume that FUDESS is entirely dependent on material configurations in the brain (or potentially other parts of the body) then swapping FUDESS between bodies is impossible. This immediately resolves the problems of swaps between neighbours and their garden spiders. However, there are other problems.

When Kirk is transported from A to B, then this certainly entails his transient extinction. At the beginning of the transport process he is dissolved and only re-constituted at the end of the process. The question now is whether or not the Kirk at the beginning and at the end of the beaming is indeed the same person in the sense of him being the same FUDESS. One thing we know is that the pre-transport and post-transport Kirks are indistinguishable to his surroundings. He retains all the memories of his life, his moods and immediate thoughts. If he was agitated before beaming, when he arrives he will also be agitated. If he felt a strong affection to some alien or human female (as he so often seems to) then he will still feel this affection after beaming. Any emotional state he was in right before being transported he will be in immediately after. So, it seems that there is a continuity of personality and of behaviour. This also implies that the person who is in Kirk's body after transporting will think that he has been Kirk for all of Kirk's life. He has the full autobiographical memory. The first day at school, the first kiss, the day when he joined the Starfleet academy and so on. So, he is fully Kirk, even though every atom that makes him is new.

It could be, however, that the transported Kirk is a completely new person. This completely new person is an exact copy of

Kirk, that is in every aspect like the old Kirk and for all the world indistinguishable from the original Kirk. Still it could be that a different FUDESS is inside Kirk. Think about the swap with your neighbour. Provided you lose your memory and gain that of the neighbour, then you would have no way of knowing that you body-swapped; you would have to assume that you have been the neighbour all the time. Similarly, the new Kirk FUDESS, which inherits the brain structure of the old Kirk FUDESS would not know whether or not it is identical with the pre-beaming Captain, or whether it may be a new FUDESS. Note that we cannot ask Kirk, because after beaming, he will certainly be under the illusion to have existed all along, because he inherited the autobiographical memory from the original FUDESS. We may throw in the towel at this stage. It seems that our question, whether or not the FUDESS before and after transporting is the same is undecideable, tethering on the verge of being completely meaningless. What good it is to speculate about undetectable effects caused by science fiction technology?

At this point I ask the reader for a little bit of patience. What seems to be utterly irrelevant and undecidable will shortly turn out to be not irrelevant at all. Before I can support my claim I have to clarify one more assumption. I assume that the FUDESS is a centralised experience in the sense that it is perceived as a central processing unit where all the conscious information comes together and is processed. Note, that I am not claiming that it is implemented in the brain as a single central processing unit, I only claim that the first person experience seems to be centralised in one mind. To use another metaphor. If the FUDESS is a window out into the world, then it is exactly one window. We cannot see or experience what is happening in Ashford, England at the same time as we are also in a body in Harbin, China. What this really means is that one FUDESS can be associated with at most one body at any one time. (Indeed, it even seems to be the case that one FUDESS can only pay attention to one particular detail at any one time.)

Coming back to Kirk let us now investigate in more detail what the transporter is doing to him. Remember that the teleportation works by reading and storing the particular atomic configurations that make Captain Kirk, it then dissolves the original and re-establishes an identical copy somewhere else. Usually in Star Trek they (quite understandably) gloss over what happens in-between dissolution and re-constitution of Kirk. This transient phase is interesting though because there must be a point between departure and arrival where the Captain only exists as information. It could be that re-constitution and dissolution overlap in time, but this does not change the conclusion. For every part of Kirk it was true that at some point, maybe just for a very short moment, it only existed as data in a computer. We may therefore as well think of dissolution and re-constitution as non-overlapping. Precisely for how long individual bits were non-existent will depend on how the teleporter works (which we will never know) and the distance between the point of departure and the destination.

The possibility to store objects, and indeed people as information and re-create them as matter at any point later offers an interesting *gedankenexperiment*: Let us imagine that the person operating the teleporter decides to break the (undoubtedly existing) rule that the scanned information of a person must not be stored. What if it was stored? With Star-Trek technology it is now possible to 3D-print as many Captain Kirks as one wishes, at any time, either a minute after the original teleportation or even years or decades later. If she wishes to do so, the operator can flood the world with copies of Kirk. Remember now that we have assumed, or rather the creators of Star-Trek have assumed that FUDESS is entirely material and that the person before and after scanning is the same. Indeed, we found above that it is very difficult to doubt this proposition, because the pre-and post-transported Kirks are completely indistinguishable, both for the world at large and for Kirk himself.

A new perspective arises when we allow the copied information to be realised more than once. Each Kirk that is created will feel that he is the only real one. Each of them would have all the autobiographical memory available to him in vivid detail. Each of them would remember the same kiss when they remember their first kisses, each of them would have the same fond memories of their time at the Starfleet academy. At the same time, it is clear that they cannot be the same essence. If you stick a needle into the arm of Kirk 23, Kirks 1-22 and Kirks 24-N will not scream. They will not feel the pain. The Kirks would not be able to read each others' thoughts any better than they could read anyone's thoughts. Two Kirk's could even fight with one another to death. In other words, they share the same autobiographical memory up to a given point, but beyond that they are independent people.

We can now also ask which of those many independent Kirks is the true continuation of the pre-teleportation Kirk? The first one maybe? Or the second one that has been made or number 23 or maybe none of them? Where does the FUDESS of all the others come from? What happened to the pre-teleported Kirk FUDESS? Is he dead?

When we asked above whether post- and pre-transported Kirk is the same FUDESS, it seemed like a meaningless question; it appeared unreasonable to doubt that they were just as it appears unreasonable to assume that ten minutes ago you were not the same person. When we expand the example to several copies, it suddenly becomes obvious that it is not that meaningless after all. Indeed, with so many Kirks milling around, it appears nearly absurd that any specific one of them is the same person as the original Kirk. All of them are valid copies with the same properties, but the feeling and experiencing pre-teleportation FUDESS probably no longer exists. The lights were switched off for good when Kirk was dissolved for the first time.

Let us now try to understand what our Star Trek *gedankenexperiment* reveals. As long as we transport Kirk alone and stick to the rules, no problem seems to occur. Everything is fine, Kirk exists, Kirk is dissolved, Kirk is re-established somewhere else. Once, we forget about the rules and we make many copies of the Captain, then we start to run into a paradox. What seemed to be a safe assumption, that transporting leaves the FUDESS unchanged, becomes difficult to accept now. You may think that the whole problem is irrelevant because transporters do not exist. However, the transporter example was just a vehicle to illustrate the problem in the assumption that FUDESS is matter and its interactions alone. If we assume that, then we run into paradoxes.

To clarify this further, let us now do another variant of the same thought experiment. Assume that you live in the far future, in a society in which science has progressed to the extent that it is actually getting to a point where scientists can make carbon copies of individual people. Exact atom for atom replicas of living objects, just as the teleporter in Star Trek can do them. The technology is now safe and ready for use, but still needs to be certified. For this they need volunteers. Assume now that the scientist in chief would pay you 2 million Pakistani Rupees if you allowed her to make such a copy of yourself as a part of the certification process. The condition is, however, that you would agree to be killed (in some humane and painless manner) just before the new copy of you goes live. So, the sequence consists in scanning you, saving the copy, killing you and then 3D printing the new atom for atom replica of you. In essence the scenario here is the same as the transporter example above, but without actual transport. Once your new you comes into existence, it will think that it has been alive all the time and feel a continuity with the past, share with you your first memories. Yet, would you really agree to this for 2 million Pakistani Rupees? Or 5 million? Or any other amount of money?

Well, you would probably if your family has fallen on real hard times and you feel the need to make the sacrifice for their sake. However, if I were you, I would think twice if your motivation is to enjoy a new life on a big yacht with your girlfriend or wife, or both of them. The Kirk thought experiment would make me worry that the lights go out for good for the pre-experiment I, the real me (or the real you in this case). The post-experiment FUDESS will then be a different person to enjoy the pleasures of newly-found wealth and yacht. The new you may well be indistinguishable from you and may even think she is you, but that is no good to the pre-copy you who lingers in the kingdom of Hades.

FUDESS and consciousness

The two models, the Star Trek scenario and the Avatar scenario roughly correspond to two different philosophical base-assumptions. The latter is compatible with the idea of a non-material soul, as some religions have it. It seems to me that it is very hard to defend. Physical changes of the brain can impact personality, memory and even cause mental disease. In the light of this it is difficult to understand what role a non-material FUDESS could still have. Moreover, the body-swap example is difficult to make sense of. There are therefore convincing arguments against the notion of a soul as a carrier of significant pieces of personality. By exclusion, we have to assume that FUDESS is a material effect, that is, it is entirely a consequence of material interactions. *Tertium non datur*. However, the Captain Kirk example shows that this view is also difficult to make sense of. If FUDESS is a purely material property, then teleporting should work. If an exact atom for atom copy of say, you, is made, then your FUDESS should transfer into that. The problem is that it if it is indeed possible to make one copy, then it is possible to make several copies. The question then is into which of the copies *your* FUDESS will transfer. Through which of the copies will *you* then see the world?

So, a material FUDESS is only defensible, if FUDESS does not include a first person experience. This, however, amounts to defining away the problem.

Before proceeding it will be useful for the rest of this chapter to clarify the connection between consciousness and FUDESS. The reader may well have assumed that what I have called FUDESS here is just a slightly odd new name for consciousness. Indeed, there is an overlap between FUDESS and consciousness. However, they are not the same. It is difficult to be precise when talking about the first person subjective experience. Theologians and Philosophers have been interested in FUDESS for millennia. They have also recognised that there is a conceptual problem related to it. More recently, neuroscience and computer science have attempted to understand consciousness from their own perspectives. There is now a respected and rather large interdisciplinary effort going on to understand what it means to be conscious and how this phenomenon comes about. Computer scientists collaborate with psychologists, physicians and neuroscientists to understand what it is that gives us this particular first person experience that seems so mysterious.

Scientific study of the private subjective experience is very difficult. FUDESS, by its very nature, is not measurable. It is subjective and therefore not accessible to the scrutiny of peers. It cannot be studied objectively because nobody except the FUDESS itself can have access to the FUDESS. This property makes FUDESS one of those things that are best left out from scientific enquiry, just as we leave out the study of the beauty of Mona Lisa or why Lars likes fishing more than skiing. It is in many ways a no-no topic for science. On the other hand, FUDESS is such a striking phenomenon that it is hard to ignore it. This is a difficult dilemma for science, but there is a way out. Scientific research into consciousness tends to simply ignore the FUDESS aspects of the consciousness phenomenon, and focuses on objectively definable and measurable aspects. Typical questions relate to the specific areas of the brain that are responsible for consciousness, objective measures of consciousness that can distinguish between sleep, wakefulness and unconsciousness based on recordings of brain activity. Another direction of research is how and when perceptions enter the conscious experience, as opposed to being merely sub-consciously registered.

Much has been found out about consciousness and much remains to be found. Despite the progress of the science of consciousness, it is not at all clear that the particular phenomena that scientists research and explain come to the core of the FUDESS puzzle. They do not explain the mystery of the first person experience. To be more precise, they may explain what is physiologically necessary for this experience to occur, but they do not explain *your* particular inner life. The FUDESS mystery is precisely the bridge between the objectively measurable events that go on in us and the subjective experience that each and every one of us perceives. In what follows, I will argue that objective scientific research does not, indeed, cannot build this bridge. This does not mean that consciousness research does not find out about consciousness, it only means that "consciousness" as defined by the practice of this science is not the same as FUDESS. In brief, consciousness research can tell us what goes on in our body as we go through the day. How various psychological effects manifest themselves in the brain and what is needed so that we have the first person experience. However, it does not explain why *you*, dear reader, have the first person experience you have, why you are in the body you are, why you exist now and not 200 years ago, whether you would have existed had a different sperm cell made the race for the ovum on the day of your conception. The problem is the *you*, the intimate experience of being in existence that is very private to you.

The main approach of research into consciousness concentrates on the brain as the presumed seat of the phenomenon.

Relatively new technologies allow scientists now to take a look at the living and thinking brain, which also enables them to pinpoint areas of the brain that are responsible for various tasks. For example, vision is processed in one part of the brain, whereas tactile sensation is processed somewhere else. With novel imaging technologies it is also possible to observe the brain live as it performs various tasks. There is now also a substantial medical literature on the effects of localised brain damages on behaviour. Computational scientists produce simulation models of the brain. Combined these allow powerful new insights about how the brain works and how the phenomena we experience are generated. Neuroscience is very good at finding so-called neural correlates of conscious actions and conscious experiences. Neural correlates are measurable brain activation patterns that coincide with a particular experience. Say, you are in love, then there will also be a specific pattern of neurons that fire. This is the neural correlate of love. It is detectable in the brain. If you are hungry it is the same. It is not inconceivable at all that one day scientists will be able to know what you are thinking, at least partially, just by looking at objective measures of your brain activity.

The holy grail of neuroscience is to find the neural correlate of consciousness, that is to identify precisely what it is (in the brain) that generates the conscious experience. Huge strides are being made in this direction. There is no reason to assume that we will not one day understand precisely where in the brain consciousness is produced, what is needed for the lights to go on.

The discovery of the neural correlate is commonly confused with a complete understanding of the phenomenon. Sometimes, particularly in the popular science literature, there are even undertones of a naive extreme neuro-reductionism. The existence of emotional states is often denied once the neural correlate is known. For example, love is *nothing* but a particular pattern of neurons firing. For anybody who has ever been in the grip of emotions of any kind, it is clear that these are quite real. If they are reduced to a neural correlate alone, then this amounts to denying the reality of the subjective experience, which is absurd. Therefore, the neural correlate of X is only a partial explanation of X and not the phenomenon itself. Indeed, at present neural correlates are often not explanations at all, but merely observations of coincidences. That is, whenever somebody is in love, then the particular neuronal pattern Y is observed. Precisely how the pattern itself is generated is not specified. Still, we have no reason to assume that it will not be possible one day to give a very detailed causal explanation of neural correlates.

Still missing from science is then the bridge from the correlate to the subjective experience to which the correlate corresponds, that is a conceptual bridge from the experience of seeing blue, or red or any other colour to the brain pattern corresponding to blue or red or whatever colour one focuses on. Without this bridge neuroscience is like a congenitally blind person who gets a perfect description of a landscape. She will know everything about the view, but she will still not actually see it. The problem is that this bridge has to link the world of the objective data with the world of the private incommunicable experience, the concrete with the unspeakable. If neuroscience one day explains consciousness in physical terms, then this hypothetical theory of consciousness will still not be conscious itself.

At present we do not have a complete theory of consciousness. Still, it is clear what such a theory would have to look like to be worth its salt. For one it would have to contain a complete understanding of the functional properties of the brain. What area does what? It would have to include a bottom-up understanding as well. Instead of merely observing what the brain is doing, we would also have to be able to build artificial brains that replicate these capabilities, at least partially. A theory would not be complete without theoretical models, that is a formalised understanding of what goes on physiologically when various signals from different parts of the body are generated, relayed and processed in the brain. So, a complete theory of consciousness and the brain would be able to understand precisely what happens when I stick a needle into your arm and why this causes pain and what goes on when you feel the pain. However, and this is the crucial point as far as FUDESS is concerned, the complete theoretical understanding of what happens when I stick a needle into your arm, would not actually cause you any pain, it would not be the experience itself, just a description. This is why a theory of consciousness is not a theory of FUDESS.

The inability of such a hypothetical theory of consciousness to cause pain is just another aspect of the Captain Kirk contradiction. When he beams from one place to the other, it is objectively indistinguishable whether or not pre-teleportation and post-teleportation Kirk are the same person. For us it makes no difference, because we have no way to distinguish the two cases. However, for the pre-teleportation Kirk it does make a difference. The transporting may well mean that his lights go out forever. Anybody with a will to live should therefore be very hesitant about teleportation. Yet, even a perfect theory of consciousness could not help us or Kirk to decide whether or not teleportation is safe, because this theory would not be able to connect scientific description of physical process related to consciousness to the subjective experience of the lights being on.

The bridging between a physical theory of consciousness and the actual first person experience is sometimes called the "hard problem of consciousness." In order for there to be such a hard problem, it has to be assumed that there is indeed something to be explained, that is there is a FUDESS. My point of view is that every one of us, who has a FUDESS, can be certain that they exist. It would be contradictory and irrational to be convinced that you do not exist. At the same time, the phenomenon of FUDESS is deeply puzzling. Within a scientific *weltanschauung* the FUDESS is an intellectual oddity. It is unmeasurable, it is even hard to talk about it in a meaningful way; it is unspeakable.

It seems that we have the hard choice. Either we have to assume that FUDESS does not exist. This requires some intellectual acrobatics and is contradictory. It would amount to convincing yourself that you do not exist. Alternatively, we could accept that FUDESS remains mysterious forever. This possibility is clearly not acceptable for SR, because it would mean that there is something that is in principle unexplainable by science. Our third option is to assume that FUDESS is non-material, that it can be transferred between bodies and that it is independent of the brain and its states. This would allow body swaps and is inconsistent, as discussed above.

Needless to say that this mystery of FUDESS, or the hard problem of consciousness, does not sit well with some of our contemporaries. It is more than an unknown, poorly understood phenomenon. It is a dent, or rather a gaping hole in our *weltanschauung* too big to cover up. It is a Kuhnian anomaly in the scientific paradigm that defines our age. It is a problem for those who like to believe in an immortal, immaterial soul. We do not need to go deep into neuroscience to know that the idea does not make any sense. At the same time, it is also a problem for those who want to believe into a world that is matter, matter and matter alone. Explaining FUDESS in terms of science alone does not work.

There are several ways in which our contemporaries react to this contradiction: Some accept that there is a problem that we cannot solve right now. However, life goes on, food needs to be put onto the table, so let us worry about other things. The second reaction is to take this mystery as a proof of the existence of God and to reject any earthly pleasure, including science, women and opera, and become a religious fundamentalist. The third option is to convince yourself that the hard problem of consciousness does not exist, that there is no such thing as FUDESS. Interestingly, while the religious people are usually quiet and peaceful, reaction 3 is vocally supported by a class of jihadists of SR. Once you start to notice them, they pop up everywhere. They are not physically violent, but they put their point across with considerable aggression. There is a hysteria that has gripped a section of society, a militant atheism that is pure emotion, an irrational response, probably caused by early childhood experiences rather than by careful thought. The hysteria is so strong that even respected intellectuals risk their name and reputation writing militant pamphlets about religion that are full of inconsistencies, uninformed and often simply poorly argued.

I do understand that some of the proponents of SR struggle with the FUDESS problem psychologically. The issue touches on one of the central tenets of SR around which they have built their lives. This can be unsettling for some. If you have defined yourself as a "woman of science," if you have lived your life in apparent rationality, fought against spiritualism and myth, denounced superstition, maybe you even have a bumper sticker with a fish on legs, then this has become a part of who you are. You define yourself, at least partially through your belief in SR. You have put all your eggs into the science basket. It has become your particular *weltanschauung*. FUDESS is a crack in the cathedral of SR. It is an unsettling problem.

Some feel they need to enter into a jihad over this issue. They want to cling on to a pure and untouched version of SR where everything that is not science simply cannot be. This desire to conserve a belief system can lead to the most fantastic acts of intellectual acrobatics aimed to hide irrationality within a flurry of seemingly rational argument. This is rarely as clearly spelled out as in the first sentences of Thomas Metzinger's book "The Ego Tunnel." He writes:

[...] I will try to convince you that there is no such thing as a self. Contrary to what people normally believe, nobody has ever *been* or *had* a self. [...T]here is no thing, no indivisible entity, that is *us*, neither in the brain nor in some metaphysical realm of the world. (Metzinger, 2009, p. 12)

What Metzinger says is quite extra-ordinary. Take a few moments to let this sink in. He claims that there is no *us*, meaning there is no *you* either. According to Metzinger, dear reader, you do not exist. Most of all, according to Metzinger, Metzinger does not exist. This is a remarkable conclusion to reach. Now, the question we have to answer is: Is the conclusion based on rational thought, or is the conclusion simply reached as a way out from the FUDESS dilemma? Is it a way to save SR and thus Metzinger's emotional comfort, or is there some substance to it?

Metzinger is not the only one who comes to this conclusion. There is a whole philosophical tradition, reaching back all the way to David Hume, that denies the existence of FUDESS, or the "us" as Metzinger calls it. The reasoning used to arrive at this conclusion is always roughly the same, although with times and scientific progress the specifics change. One version is just the Captain Kirk argument from above. The basic structure of the argument was as follows:

1 We can copy Captain Kirk using the transporter facilities.
2 Each copy of Captain Kirk will be exactly identical to all the others.
3 Still each of the copies will believe that it has a unique self/I/FUDESS.
4 Hence, to avoid a contradiction, we must conclude that the self/I/FUDESS does not exist.

This argument seems to be very convincing at first, however, a closer look shows that it is also circular. In order to get to step 4 from the premises, we already had to assume that anything that is not measurable or otherwise detectable using the means of natural science is disallowed. Now, FUDESS is subjective experience and as such not accessible to direct scientific scrutiny.

Hence, all the above argument really says is that we cannot measure what we cannot measure, which is trivially true.

Believers of SR and deniers of FUDESS tend to argue not with the help of *gedankenexperiments* borrowed from science fiction series, but instead they base their arguments on research results, particularly from neuroscience. Metzinger is a case in point. To some extent this is a very valuable strategy. Philosophers all too often talk about science without really understanding science, which makes bad philosophy. However, there is also a deep problem with basing philosophical entirely on empirical research: Science itself comes with its own philosophical baggage. It is not a value-free, philosophy free activity, but relies on strong assumptions about what the world is like and how it can be discovered, what is a valid argument and what is and is not an interesting research problem or result. For a philosopher there is always the danger that accepting empirical results also means getting a few cuckoo eggs of SR into her nest. If philosophers like Metzinger are so convinced of the value of neuroscience for philosophy, what sorts of results are they looking at? What is it that convinces them that there is no self? Or at least what is it they use when they make their argument?

One of the basic observations that comes from neuroscience is the architecture of the brain itself. It appears that the brain, as a system, is highly de-centralised. Naively, one tends to think that there must be, somewhere in the brain a centre in which the FUDESS sits. The US philosopher Daniel Dennett calls this the Cartesian Theatre. The idea that there is a well localised place in the brain where it all comes together, where consciousness is happening, so to speak. It becomes increasingly clear that this is not the case. There is no such theatre, at least nobody has found it yet. Instead the brain seems to be distributed without any centralised command and control structure.

The lack of such a centralised self may come as a bit of a blow for those who firmly believe in the real existence of a "self." If you cannot find the self in the brain, then it probably does not exist. On the other hand, maybe it would be unreasonable to expect to find a well localised me. Imagine for a moment that scientists indeed discovered such a localised self in the brain. A well defined area where FUDESS is produced. What now? Certainly, scientists would want to know *how* the self works. So, the first thing they would do is to try to take it apart and look how it works, just as Monod took apart his *E.coli*. No scientist would be satisfied to know that there is a Cartesian Theatre, but they would want to find out how the curtain works, where the stage-doors are, who the audience is and how they get into the theatre. They would want to disassemble the theatre into flat-packs. This is what scientists do. Soon, the initial sensation of joy at the discovery of the self would give way to a dissatisfaction with the lack of understanding of the inner workings of the newly discovered self. As science learned more about the self, it would notice that far from being a simple atomic entity it is a distributed system. Its stable operation depends on many parts working together. It has to. If there were such a thing as a FUDESS atom, a single particle that produces FUDESS but cannot be explained itself, that would be nothing short of mysterious. It would be akin a spirit particle and contravene everything we think is possible. The closer we look at any putative Cartesian Theatre, the more we take it apart into flat-packs, the more it evaporates in front of our eyes and we would fall into the information black-hole again.

The conclusion to draw from this is that the idea of a centralised seat of consciousness or FUDESS will be frustrated anyway, no matter what the architecture of the brain. In a sense, the brain *is* a centralised seat of consciousness when it is considered at an appropriate scale. Not surprisingly though, as we drill down the hierarchical layers of organisation the centralised system becomes quite de-centralised.

Many philosophers of consciousness seem to be interested in

research into the degrees of consciousness. At first it may seem that there are essentially two states of consciousness. Either the lights are switched on, or not. A closer look reveals, however, that there is a continuum between being fully conscious and completely unconscious. Quite trivially many tasks we perform are done with little or no conscious attention at all. A near classic example is driving back home from work, or any other very familiar route. The driver does not need to pay particular attention to what she is doing, how she changes gear, where she takes a turn and so on. All this can happen while listening to the radio or thinking about something completely unrelated. The driver will clearly have some conscious attention on the task, but much less than if she were to, say, solve a crossword puzzle that would require complete attention.

Other things we do are even less under conscious control and probably even token-voluntary in the sense that we *could* control them if we only were aware of them. Just observe, during a meeting, what the other participants are doing with their hands arms and legs. Some may tear up little pieces of paper, others probably play with their pen, some may twitch their legs, they may scratch themselves, play with their hair or anything like that. It would be perfectly possible for the person to control her movements, if she wanted to. However, most of the time, she is probably not aware of them.

Related to this is a famous experimental setup to demonstrate change blindness. A typical experiment involves subjects being recruited to some decoy-experiment. As a preparation for this they are asked to fill in some forms to provide basic consents and information about themselves. The relevant forms are handed to them by a research assistant posing as a receptionist. The subjects then fill in the form and hand it back to the assistant. Only that at some point during the process the assistant was discretely switched to the effect that the subjects handed back the form to a different receptionist/research assistant. As long as this swap is performed discretely, the subjects rarely notice the change, even if the two assistants look very different from one another. What this tells us is that we are not fully conscious all the time but partially switch off quite often. We also do not have an unbiased view of the world, but instead we are very selective in what we consciously perceive and what is left unnoticed.

These phenomena are very interesting and they also tell us something relevant about how we and our brains work. There is a very large body of research now that probes into how we perceive the outside world. Consciousness, the argument very often seems to say, is not so much the God-given clear view on the world she has created, but rather a highly selective representation of the world, an evolutionary trade-off that allows us to navigate more efficiently to potential mates. One conclusion to draw from this is that we are robots that are most of the time on auto-pilot. Consciousness is but a small part of the experience of the world we have and it only contains the most essential aspects of the world.

One should always keep some healthy scepticism when interpreting research results. Effects such as change blindness are astonishing, but ultimately they have been obtained in very particular experimental set-ups. They certainly tell us something about consciousness, but how relevant they really are in daily life is a different question that still needs addressing. In any case, the real question is whether the fact that consciousness comes in degrees is relevant for our understanding of FUDESS. I doubt it is. It may well be the case that we are partially switched off at times, or we focus on one particular narrow aspect of the world forgetting the rest around us. Still, we experience ourselves during those times; there is a FUDESS. Even if you are hallucinating or drunk. There still is this "thing" that has a warped perception of the world. The experience of FUDESS, the first person experience, does not go away. Whether what we experience is a lie, or an illusion, whether what we experience is but a small part of the world, we still have the first person experience.

Consciousness research in psychology and neuroscience addresses important objective aspects of consciousness. It completely leaves out the subjective experience, the FUDESS, the perceiving entity. Psychologists try to understand what precisely it is that is perceived. Neuroscience can understand what is going on in the brain of the perceiver, as she perceives. Yet, neither of those can explain to you and me why it is me or you who perceives, how the objective insights about consciousness are translated into pain, emotion, joy or simply just the sensation of being that each individual has. Science is limited to studying the objective manifestations of these emotions, that is the nerve signals, the neural correlates, your reactions to them and so on. All these manifestations, however, are not identical to the experience of the emotions themselves.

Another aspect that is often thought to be important is the self-model aspect of consciousness and the control over a body. As I am typing these lines, I am perceiving my hands, I feel the resistance of the keys against my finger tips. We all think of certain parts of the world as belonging to us and most of all we feel these parts. This "part of the world" is primarily our body, but not exclusively. A famous experiment to probe the relationship between us and the body is the celebrated rubber-hand experiment. The set-up is very simple. The subject hides her hand from her own view, for example by putting it under the table in front of her. An experimenter then strokes her hand. At the same time, the subject has in full view of herself a rubber hand that is stroked as well, in the same way as her own hand is. After a certain amount of time observing the rubber hand and feeling the stroking sensation, the subject will feel that the rubber hand is her own (rather than her proper hand that is hidden from view). A simpler and less dramatic effect is probably known to most of us. Expert users of tools will, after some time, start to perceive the tool as an extension of their own bodies. They can feel and sense using this tool. The classic example is a blind person who uses her stick to "see" the world. After a long time of use the stick will become an extension of the user's body. Computer gamers will often feel that their joysticks or whatever they use becomes an extension of their body and the virtual figure they use an extension of themselves.

There are again certain diseases that lead to a disruption of the body perception. Very specific damages in the brain can lead to patients feeling that parts of their body "do not belong to them." Again, all this is interesting from a research point of view. However, it is not of primary importance for our understanding of FUDESS. Even the person who thinks that her arm is not part of herself, still thinks, still is. While her dysfunction is clearly associated with consciousness, it leaves the FUDESS itself unaffected. We can still ask: Who is this person who thinks that her arm does not belong to her? Similarly, the rubber hand experiment probably does not get us any closer to understanding FUDESS either. It is a phenomenon, apparently striking to those who have experienced it, but it does not touch on the question of FUDESS. Who is it that thinks that the rubber hand is hers?

This conclusion generalises to almost any area of consciousness research. Science always re-defines the consciousness problem so as to shut out the question of FUDESS. It has to be like this. A science of FUDESS would probably not be a good science. The subjective cannot be phrased in objective terms. In order to study consciousness, FUDESS has to be left out.

The brain as a computer

Every age has its own limited range of metaphors and experiences that it can use to understand phenomena. Among the proponents of SR there often seems to be an undertone that we can do pure science, that good scientists can detach themselves from their own biases and prejudices to concentrate on facts and data alone. Of course they cannot. Not only is it misguided to think that anybody can view the world without prejudice, it probably is the case that we have to have pre-conceived ideas about nature in order to do science in the first place. Data itself is meaningless. A scientist cannot learn about nature unless she is able to fit her results into a wider context of knowledge and understanding.

When it comes to the brain a common pre-conceived idea is to see the brain as a computer. Indeed, much of the contemporary debate about the brain and its function overtly or tacitly assumes that the brain is in essence a computational device, a biological information processing machine. To some extent this is a fruitful way of looking at it because it allows us to ask interesting and fruitful questions. Not only that, thinking of the brain as performing computations can and has also given rise to novel methods in computing. Artificial neural networks are a class of algorithms that are directly inspired by the architecture of the brain and find widespread use in artificial intelligence and pattern recognition in computer science. Useful as the brain-as-computer model may be, it also has its difficulties.

The brain performs many computational tasks, but it seems to do more than that. When we think of computers, these days, then we think of entities that can be used to perform text processing, watch movies and to display graphics. The screen is linked to our mental image of a computer to the extent that we find it difficult to separate the two. In reality, graphical peripherals are fancy accessories. The computer at its very core is a number crunching device. It has no colours or pictures. It is not that long ago that the only way to interact with computers was via punchcards. Long stripes of paper with holes in it encoding programs and their output. There were no screens to represent the inner state of the machine.

When we think of the brain as a computer, we need to think of it as a processing unit in separation of its peripherals, specifically, we need to imagine it as being without a display. The brain, as it is, is just a computational device that can receive inputs and generate outputs, but by itself does not have a screen or anything like that. We can think of it as a micro-controller like the ones that can be found in autonomous robots or the Visilab car. A computational core that takes input and produces some output. It is the processing device that controls the body, generates internal states (i.e. "hungry") or calculates and controls the execution of appropriate actions (i.e. "run away").

From a biological perspective, this is all that is required, a micro-controller that processes input data to generate the appropriate output. One difficulty of the computer metaphor is that the brain seems to be more than that. It actually does have a screen attached to it. Not in the literal sense, but it has something that makes the world show up. It is able to make us know what sounds sound like and views look like, foods taste like and surfaces feel like. It has a multi-output user-interface. Bearing in mind that a user-interface entails a user we can ask: Who is the user? Who is it that operates the computer? Who experiences the output from the computer? Well, you are! And here is the problem: You are not really. It is impossible for you to experience anything without the processing power of the brain. All your sensory world is brain/computer generated. You can only observe because you have a brain that does the observing for you. In this sense, there is no independent you in this picture. The computer observes itself, it is its own user.

There seems to be a screen somewhere within the brain that we watch. At the same time, if we and our experiences are but the brain and its computational power, then we also generate the pictures that we are watching. On the other hand, if the brain was but a computer, it would not need a screen. There would be no need for the computer to interface with itself. It could simply process inputs and provide outputs, just like the micro-controllers of robots. Machines do not have and do not need an internal life. The Visilab car does not need a screen to navigate to China.

This does not mean that there is no need for representations of the world. The Visilab car has, at any one time, an internal representation of itself and its environment. However, as far as we know, this representation is not experienced by the Visilab car itself. There is no sensible way to say that the car is aware of its place in the world. You could think that we bio-machines could function in the same way, as processor controlled units without any need for a FUDESS. Yet, apparently, paradoxically, quite mysteriously, there is an observer in there: Me, in my case, you in your case. This observer is quite undetectable by measurement, even conceptually problematic. What is more, the observer and the observed are the same thing, or at least overlapping. The brain observes the brain generated illusions of a world.

Seen from this perspective the notion of FUDESS or self is paradoxical. As long as you think in terms of a machine observing and representing itself, you cannot conceptually derive an independent self from it. This independent self would have to be somehow different from the computer that generates the illusions. It would have to be some sort of external screen and screen observer. Yet, this observer would itself need a brain to process and understand the observed. So, we move in an infinite loop.

There are two possible conclusions to draw from this. Either we give up the notion of the FUDESS. There is no self, just as Metzinger said. Then the problem disapears, incidentally along with the Captain Kirk paradox, but you generate another one, namely that you have to deny your own existence. The second choice is that you abandon the view that brains are just computers. In this case the Captain Kirk problem remains and we still do not know what brains could be over and beyond common processing devices.

Digging deeper

Notwithstanding the conceptual problems, let us assume that the brain is a computer. At the most basic level the brain can be conceptualised as interconnected neurons. Each of these neurons receives inputs from one or more other neurons. Essentially, this input can be either on or off. When a neuron receives signals from other neurons then it may activate itself and in turn send a signal to whatever it is connected to. Some of the neurons in the brain are connected to the sensory peripherals of the body, the eyes, the ears and touch senses and so on. Others are just connected to other neurons in the brain itself. Collectively the neurons can process inputs from the environment and produce some output in the form of control of various bodily functions including motion, speech, breathing and so on.

Each of those neurons is rather complicated in its internal structure and function. Still, there is no reason to assume that we cannot comprehend and describe what it is doing. We could, for example, try to model neurons using computers, that is we could write computer programs that mimic the connections of neurons and their function. If we just had enough time and patience, then we could probably replicate an entire brain. This is a very difficult enterprise, but there is no reason to believe that it cannot be done. Maybe this will not be done today, maybe not tomorrow, but not too far in the future somebody will be able to do it.

Indeed, there are attempts on the way to do just this. One is the Blue Brain Project at EPFL in Lausanne, Switzerland[5]. The Blue Brain team uses super-computers to build artificial brains. The challenges in terms of required computing power and modelling skill are enormous. Consequently the beginnings are modest. For the time being the best they can do is to model a small part of a rat's brain. Building an entire human brain *in silico* is still quite a long way away. The challenge is not only to get the raw computing power, but also to specify the right connections between neurons. At the moment we simply do not have enough detailed information about how precisely brains are wired up. Sophisticated future brain imaging methods may need to be combined with bottom up constructed brains to make simulated copies of real brains.

This is science fiction for now, but probably soon it will be reality. When the day comes when the first simulated human brain becomes virtual reality, then we will face an interesting philosophical problem. Will this first simulated brain have FUDESS? If so, who is experiencing? Where does this person come from? Are we cruel to keep it (him? her?) locked into a computing device with no means of locomotion, no way to make experiences in the real world? Or is a simulated brain somehow different from real brains, not entitled to the moral rights that are afforded to flesh and blood versions of us?

Strangely enough, even though we will (almost) by definition know everything there is to know about this brain, we will still not be able to look into it. We would still remain uncertain about its inner life, what it feels, indeed whether it feels anything at all. Even with complete knowledge the FUDESS has to remain unobservable. Clearly, any future simulated brain if it is worth its salt would have to behave in interesting ways. It will have to be able to communicate with its surroundings and to be able to hold up a conversation, probably also navigate autonomously and maybe even display some creativity. We would expect that from a brain. The question is whether or not this machine will be just a hyper-sophisticated version of Eliza, the artificial Therapist developed in the 1960s at MIT, or whether it will be a conscious and aware entity with moral rights.

The first thing that often comes to mind to decide those things is the so-called *Turing Test*. The idea is simple. In order to decide whether or not a machine or a computer program is intelligent, we check whether it can be distinguished from a real human. To do this, we ask a tester to have a remote conversation with the machine and a human control. After the conversation, we then ask the tester to guess which one was human and which one was a machine. Truly intelligent machines, so the idea, should be indistinguishable from the human. The Turing Test is very famous, but not universally accepted as a relevant test for intelligence. There are huge controversies within the field of Artificial Intelligence but also philosophy about it. As far as we are concerned here, the Turing Test is irrelevant. It is a test to assess the capability of a machine, not its inner life. The Turing Test does not tell us anything about FUDESS.

Would we have any good reason to assume that any future artificial brain has a FUDESS. Certainly, if the machine is as convincing as Commander Data in Star Trek – The Next Generation, we will quite automatically assume that it does. We have to remain agnostic about whether or not Data or his real future instantiation will actually have a FUDESS. The question is unanswerable. Indeed, we cannot even be certain about one another. Ultimately, I only know about my own FUDESS. On the other hand, if we had a conversation with Commander Data, then it would be very difficult indeed to avoid the conclusion that he had a FUDESS, just as I assume the same about all humans I am interacting with. Maybe the question is not whether we have good reasons to assume that Data does have a FUDESS, but whether there is a good reason to assume that he does not. If the Blue Brain project eventually comes up with a fully simulated brain, I would not know how to defend the argument that this simulated brain was different from biological brains. It is made from a different material, but as long as it implements the same functions, it is hard to see why it should be any different from a flesh-and blood brain. Why should one have this mysterious FUDESS but the other one not, when indeed they are computationally equivalent? The default conclusion therefore, has to be that artificial brains have a FUDESS and consequently FUDESS is purely material.

It seems that the question of FUDESS leads to contradictory

conclusions. Depending on the approach we take, we find ourselves forced to deny a purely material FUDESS, deny FUDESS altogether or deny a non-material FUDESS. Perhaps the problem is the basic assumption we make. It is well possible that we look at the problem from the wrong angle, with the wrong conceptual toolset. While we do not know what the right set of tools would be, we can at least look in detail at the one we have and get a better feeling for why it might be the wrong one.

Turing machines

When it comes to the brain and computers, then the key thing to understand is the notion of a "computation." Computation, as it is understood in computer science, is fundamentally only a transformation of symbols according to some pre-specified rules. The basic model of computation is the Turing machine. This is not a real machine, but a mathematical construct that was invented by the British mathematician and genius Alan Turing in the 1930s. In essence it consists of two parts. A long tape consisting of cells in which the machine can write symbols. These symbols can be anything you like. They could be numbers, they could be letters, words, Chinese characters, or just abstract symbols such as a star or anything else. The important thing though is that there is a fixed number of symbols. When a Turing machine is constructed, the user must specify which symbols are used and from then on they cannot be changed any more.

The tape is usually thought of as having a beginning, but is imagined to be infinitely long. We can think of it roughly as the equivalent of the memory of the "computer." At the beginning, before the Turing machine is switched on the tape needs to be prepared with an input sequence. It may be empty, but typically the initial sequence on the tape will specify some input pattern. The input defines what is to be computed by the Turing machine.

The second part of the Turing machine is the reading head. It does all the work in the Turing machine. It moves around on the tape and it writes and overwrites symbols into the individual cells of the tape. A crucial property of the reading head is that at any time it can be in a number of states. The number of distinct states of the machine is arbitrary, but just like the symbol set of the input tape, it must be fixed for all times once the machine is designed. The simplest case is a machine with only two states. One can think of this internal state as represented by a light at the outside of the machine. If the machine is in state 0 then the light goes on. Otherwise it is off. In a machine with three states the light would have two colours, say white and red. In practice nobody goes to the trouble to represent the internal states of Turing machines by lights. Instead, they are simply represented as numbers. So, a reading head can be in state 0 or 1 or 129 or anything that is allowed by the rules.

At the heart of the operation of the Turing machine is that the reading head changes its state. It does not do so randomly, but according to fixed rules. These rules are given by the user and constitute the "program" of the machine. Every rule is of the same form:

If the head has read a particular symbol on the tape and if it is in a specific state, then it will go into a specified new state, write a new symbol onto the tape and then move to the next cell on the tape, either to the left or to the right.

The Turing machine finishes its operation when it is either instructed by its rules to go into the halting state or when it simply runs out of rules to follow, that is there is no rule that is applicable to the particular combination of tape symbol and internal state.

An example might clarify this: Imagine a very simple Turing machine, let us call it the mood TM, whose reading head has two states, *happy* and *unhappy* and the tape has three different symbols, namely *blank*, ☺ and ☹. We can now write a simple program, that specifies what the Turing machine should do. Let us assume that the reading head, whenever it encounters a ☺, changes its state to happy and whenever it encounters a ☹ changes its state to unhappy. In both cases we assume that it moves to the right on the tape and deletes the symbol it has just encountered. Let us further assume that when the reading head encounters a blank, then it writes a ☺ or ☹ if it is in state happy or unhappy respectively and halts.

Formally, this very simple program of the mood TM can be written down using a few simple rules.

state	**tape**	**next state**	**new symbol**	**movement**
happy	blank	HALT	☺	⇒
unhappy	blank	HALT	☹	⇒
happy	☺	happy	blank	⇒
happy	☹	unhappy	blank	⇒
unhappy	☺	happy	blank	⇒
unhappy	☹	unhappy	blank	⇒

The way to read these rules is quite straightforward. The first entry gives the state of the reading head, followed by the symbol that is encountered on the tape. The first line applies to a reading head that is in state happy and encounters a blank cell. It will then go into the state HALT and write a ☺ onto the tape. The last symbol ⇒ indicates that the reading head will next read the state in the cell directly to the right. Another possibility would have been the symbol ⇐ which indicates that the next cell to be read is the one immediately to the left. In the first line the indication of the next movement is clearly irrelevant because the reading head has gone into the halting state which signifies that no further action is taken.

Turing machines are theoretically interesting, because they define the concept of computation. Turing conjectured that for any computation that can be performed, there is a Turing machine, that is a particular program for the head to change state and a particular set of allowed symbols that can do this very same computation. So, if you use you computer to calculate the square root of 347, then there is also a Turing machine with suitable states and transition rules that can do exactly the same computation. The same is true for any computation you perform.

The mood TM is not a particularly interesting example of a Turing machine. However, because it is a TM it computes something, even though we are not particularly interested in the computational result. One could perhaps view it as a simple model of a person's mood, but certainly not a very realistic one. It could well be that, apart from the interpretation we have given to the mood machine, there may also be another way to interpret it. In any case, what is important to note is that the symbols that are used in the TM have no intrinsic meaning. In the above example I used ☺ and ☹ in order to make the example a bit more intuitive. However, I could have used · and ◁ or any other symbol instead. It is entirely up to the programmer to determine what she would like to use. The TM does not care about this, all it does is to follow rules, without any thought, without any regard for what it is doing. The symbols of a Turing machine are just that. Meaningless symbols. Computation is nothing but the manipulation of symbols.

An example of a Turing machine that computes something useful is given in table 6.1. Unlike the mood TM this machine has a very clear interpretation. If it is fed appropriate input, namely a properly formatted sequence of zero or more ※ followed by zero or more II again followed by empty cells, then it can determine which one of those two symbols occurs more frequently. An example of a valid input would be a tape like this:

※ ※ ※ ※ ※ ※ II II II II □ □ □ □....

Here □ stands for a blank on the tape.

The state transition table may at a first glance look quite arbitrary and intimidating. The idea of the program is very simple though. The ultimate aim of this Turing machine is to halt in a state where the first symbol on the tape indicates the symbol that was most frequently on the input. If both were equally frequent, then the first symbol on the tape should be the symbol T.

The reading head is assumed to start at the first tape entry. If this is a symbol ll, then it is clear that there are no ※ on the tape and we can be certain that it is the most frequent symbol. The computation can halt at this stage. If it is ※, then the machine needs to do more work. It will replace the first symbol by a termination symbol, T. The reading head then goes to the right until it finds the first □. Once it has done this, it then goes to the left and deletes the first ll, that is it replaces it by □. It then continues to the left until it finds the termination symbol T, which indicates that the reading head has reached the end of the tape. It then moves to the right and replaces the first ※ (if there is one) with a D. It then continues going to the right until it finds the first blank. As before, it then moves back left and deletes the first ll, going back to left, deletes the first ※ and so on. It continues to delete symbols in this way until there is only one type of symbol left. The procedure ensures that in every round always exactly the same number of ※ and ll are deleted. Hence, if after one round only one type of symbol is left, then it must be the more frequent one. If the machine finds this to be the case, then it goes into a cosmetic mode, clearing up all the D symbols it finds and replaces the T symbol by the appropriate entry and halts.

state	tape	next state	new symbol	movement
1	※	2	T	⇒
1	॥	H	॥	⇐
2	※	2	※	⇒
2	॥	2	॥	⇒
2	ƀ	3	ƀ	⇐
2	※	3	ƀ	⇐
3	※	3	ƀ	⇐
3	D	3	ƀ	⇐
3	T	H	※	⇒
3	॥	4	ƀ	⇐
4	॥	4	॥	⇐
4	※	4	※	⇐
4	T	5	T	⇒
4	D	5	D	⇒
5	※	2	D	⇒
5	॥	5	ƀ	⇐
5	D	5	ƀ	⇐
5	T	H	॥	⇒

Table 6.1: A program to determine the whether there are more ※ or ॥ on the tape. The program assumes (and only works if) all ॥ come first followed by ※. I borrowed the syntax here from Karen Britton's Turing machine simulator at http://ironphoenix.org/tril/tm.

This Turing machine performs a computation in the sense that it tells us something very specific about the input pattern, namely which of the symbols is more frequent. To be precise, it is not really doing that, but does something slightly different: It performs a number of rewrites so that it will halt in a state where the first cell on the tape contains either of the two symbols or the letter T. An inspection of the rules will show that it is always the more frequent of the two symbols that ends up in the first cell of the tape. If both are equally frequent, then the letter T will be written into the first entry. It is now an interpretation of this observed behaviour that the machine determines which symbol is more frequent. The machine itself does not determine anything. It does not "know" in any deep sense which one of the symbols was more frequent. Indeed, it does not even remember the initial input once it has halted.

What is true for our example TM, is true for any kind of computation. Computing is nothing but mindless symbol transformation. That's it. In light of this it is hard to see how one can conceptualise brains as a computers. Brains are meaning machines, whereas Turing machines are purely syntactic. Whatever they do, they always need an external observer who makes sense of their results. If in the comparison TM there is a T in the first cell of the tape when the machine halts, then this is just the result of the intrinsic rules of the particular TM, but does not mean anything unless we assign a meaning to it. When we say that a process X is a computation then then this has two possible interpretations: Either the statement is completely undefined and has no significance or it means that there is a Turing machine that is formally equivalent to the process. There is no alternative. When we talk about a computation then we should always imagine the squeaking sound of the Turing machine as it slides along the tape and erases and writes symbols. Over and over again until it halts eventually. We have to think of computation as a mathematical exercise, move left, write, move left, write, move right, write and so on.

Is the brain a computational system? Certainly, there are computational processes going on in the brain. We can also model parts of the brain using computers. In this sense the brain certainly is a computational system. The question is whether the brain is more than just a computer or whether it can be completely reduced to a Turing Machine.

There are a number of arguments for the idea that the brain is just a computer. The simplest argument is the "lack of imagination" argument: If the brain is not a computer, what else could it be? There does not seem to be anything capable of generating cognitive abilities, except for computations. This argument is not particularly inspiring, clearly. It is just a verbal shoulder shrug.

A more positive contribution is expressed by David Chalmers:

[...] the brain can in principle be described as following a set of iron-clad rules: namely, the laws of physics. On the highest level, the human mind seems extremely flexible, producing the very antithesis of rule-following behavior; yet at the bottom level, it is made up of a physical substrate, consisting of such entities as elementary particles and electrical charges, whose actions are determined by the laws of physics. Of course, in our current state of knowledge we do not have complete knowledge of these laws, but it is a tenet of modern science that a complete set of such laws exists. (Chalmers, 1992)

The argument he makes here is that the brain is ultimately a computation because it must follow the physical laws which are, so he claims, just a set of rules. The reader may note that Chalmers here needs to assume that the scientific description of the world, particularly physics, is a complete description, which is in essence the assumption of SR and may or may not be true. So, again, the argument is rather inconclusive.

There are arguments to the contrary. For example, the American philosopher Alva Noë (Noe, 2010), would argue that while the brain may be a computational device, consciousness or FUDESS are not reducible to brain processes alone. The moment you take the brain out of a body and put it into a glass jar, so the argument goes, much of what we call consciousness would disappear. It may well be that the brain is a computer, in the strict sense of the word, but the computer could not by itself produce a FUDESS. For this to appear some sort of embodiment and active interaction with the world is necessary. He does not offer any proof for his idea, but it is mildly convincing if one is sympathetic to begin with. If one decides to follow Noë's reasoning, a clear implication would be that FUDESS could not be reduced to a computing device, which avoids many of the paradoxs that we have encountered above.

Another *ansatz* is the observation that there is no strict separation between hardware and software in the brain. The neurons are clearly hardware, but the way they are connected to one another specifies the program, or the software. This is different from classical desktop computers where this separation is very strict. Whether or not this is relevant with respect to our question is not clear, but it may be. The analogy with DNA may help us to keep an open mind. Naively DNA is often described as a sort of tape that keeps genetic information. While this description is correct, DNA is also a biologically active molecule to which proteins can chemically bind. This binding activity turns out to be crucial to the biological function in that it regulates which genes are expressed and which ones are not. Similarly, the DNA is a physical structure that can fold into 3-D shapes. Depending on how precisely it folds it may hide or expose certain genes and thus influence the expression of genes. This is not to say that the behaviour of the DNA is uncomputable (it is not), but that we would limit our understanding of the system if we overemphasised a particular analogy.

Notwithstanding these uncertainties, most of the theoretical interest in brain studies and consciousness treats the brain as a computational system. So, the assumption is that the brain is a computational device and the rest of the body is some sort of interface that can collect information from the environment and interact with it. In concrete terms this means that the brain is equivalent in its capabilities to a Turing machine. From a theoretical point of view, therefore, the brain is equivalent to a reading head that is going left and right while reading and writing meaningless symbols.

We know that computational systems are quite powerful. When they do the right kind of symbol manipulations, then they can trade independently on the stock-market (and make money), they can guide a space-craft into orbit, they can encode music and films, they can even symbolically represent themselves. Yet, can they really be conscious? Can they have a FUDESS? Certainly, a general purpose computer can be used to implement a model of a Turing machine, and hence a Turing machine can be used to implement a model of a Turing machine. A general purpose computer could also model its own specific hardware and hence model itself. In a way this is encouraging. Self-representation is an important aspect of consciousness. However, when a computer represents a model of itself, does that automatically mean that it is conscious of itself? Hardly. It just means that there is a sequence of symbols in the machine that can be interpreted as a representation of a real object.

If we ask whether FUDESS can be reduced to brains and brains to Turing machines, we have to ask: Can a Turing machine be conscious in the sense of having a FUDESS? It is hard to see how. Where would the consciousness reside? Maybe in the tape of the Turing machine? That would not make any sense. We can think of the tape being made of paper and paper clearly cannot have a FUDESS. Or is it in the reading head? Or is it somehow mysteriously in both in some distributed way. None of this makes sense. It seems that reducing FUDESS to brains and brains to a computational processes leads to the reductionist information black hole. When we try to pin down what precisely it could mean for a brain to compute, then we end up with a purely syntactic symbol manipulator as a model for FUDESS. In this symbol manipulator there is no FUDESS to be found. The phenomenon we are after crumbles between our fingers and disappears.

The reader may object here and point out that the Turing machine comparison is somewhat misleading because the computers and brain receive constant input, which is not adequately represented by Turing machines. There is a certain amount of validity to this objection. However, this does not at all invalidate the Turing model. A look under the hood of a computer program will clarify this. Any large computer program is, in practice, programmed by specifying a number of smaller sub-routines that perform specific, usually quite small tasks.

These sub-routines are usually given some input, and based on this perform some basic computation that then in turn feeds back into the program at large. As such, these sub-routines roughly correspond to a particular Turing machine and the input provided to them would be the tape. Some of the sub-routines may also receive external input from sensors or users. It is no problem at all to conceptualise those as new input tapes as well.

Another objection one may have is that there are apparently examples of computers that can behave autonomously. The Visilab car is one example. It could, in principle, make its way from Milan to China completely independently of human assistance. At least, the Visilab car suggests that soon a technology will be available that allows engineered entities to be autonomous and interact with their environment in sensible ways. If we extrapolate this into the future, then it is certainly reasonable to assume that at some point we will be able to create a being that is behaviourally indistinguishable from us. Would we then not have to conclude that it has consciousness and FUDESS, even though it may be just a Turing machine under the hood?

This sort of argument is frequently made. Over the last 50 years or so there has been a very rapid development of information technology and its applications. It is tempting to simply extrapolate this into the future and assume that soon we will be able to build a Commander Data. Clearly, there is no rational reason to justify that the speed of developments will remain unchanged.

However, even if we for the sake of the argument assume that soon we are able to build a robot with the ability of Commander

Data, this does not mean much. Even though the machine may be behaviourally indistinguishable from normal humans, this still does not mean that it will also have a FUDESS. It may have consciousness in the operational sense. Maybe it will be able to divert its attention to specific tasks while ignoring others. It may be able to learn from experience. It may even be able to simulate affection and emotions. However, this does not at all allow the conclusion that it also has an inner life, a true first person experience, a FUDESS. It may, or may not. We will never be able to find out, as indeed, we are never able to be certain about each other's FUDESS.

One could argue that at this point the distinction between a true FUDESS and a simulated consciousness, where an agent behaves as though it had a consciousness while it does not really have one, becomes academic, in the bad sense of the word. If we cannot even be certain about the FUDESS of our human flesh and blood opposites, is it not an irrelevant question whether or not the robot has FUDESS, as long as it acts like a fully conscious being?

To understand this better, let us perform a *gedankenexperiment*. For the sake of the argument, let us accept that the android does indeed have FUDESS. It is impossible on principle to ascertain this, but still, let us takes the God's eye view and assume that we know for sure. For our *gedankenexperiment* let us ask Commander Data or whatever the name of the android may be to go into a dark room and sit down. We assume that this room is completely sound insulated and there is no light, nor anything else to distract the android. Nobody to interact with, nothing for it to do. If the android is built well enough and it had an inner life before it went into the dark room, then it will continue to have one in the room. It will then probably start to reflect upon things that it has experienced. Maybe it will think about some theoretical problems, or just fantasise about android-reproduction methods. Whatever they are these thought processes are completely transparent to us in the sense that we can record them. We could, for example, remotely monitor them by having the android send regular updates via the internet. We could also program the robot to record an accurate trace of everything it does and thinks. There are many possibilities to record what computers are doing even if they are not directly observed. However we do it, we can know exactly what the machine was thinking about just by monitoring. The thoughts themselves, however, are not the FUDESS. They may or may not be an objective manifestation of the FUDESS. Yet, FUDESS proper is something else. It is the subjective experience of the android, its personal experience. The truly felt "me" in the robot. This is an unmeasurable thing, but as we said, we assume it is there.

By assumption, this android is just a computer and completely determined by the algorithms that it runs and the particular states in which it is at any time. This means that we can simply take a copy of the inner state of the android, turn it off, maybe even disassemble it. If we re-assemble and re-boot the machine again, then it should just re-start its inner life from wherever it left off as long as we saved the inner state before re-boot. In fact most of us do this kind of thing very frequently when we hibernate our computers. The next day when we bring it back from hibernation then the computer will start off exactly from the point where we originally left it. So, hibernating computers and waking them again is a simple thing to do.

For our *gedankenexperiment* let us imagine that we do something slightly different. Instead of re-awakening the android in the same hardware in which we hibernated it originally, let us assume that we change the processor. We give it a slightly slower processor. It will still re-awaken in the same state, but now with the new hardware its thought processes will be somewhat slower. It can still do all the computation it has done before, but it computes at a lower clock-speed, say. We can now repeat this several times, each time we replace the original processor with a slower model. Until we get to a point where the computation is, say, a million times slower than in the original android. Its inner life, its FUDESS, now works at an extremely slow pace. As far as the android is concerned, it makes no difference. If it had an inner life before, and a FUDESS, then it still has one.

Note that changing the speed of a computation does not change the computation. If you use a computer for any computational problem, say to for a prime number factorization, then you can do this on a fast computer or a slow computer. The result will be the same on both, even though the speed with which you receive the result may be radically different. Hence, the android and his alleged inner life in the dark room should be no different no matter what the speed. It may be that it experiences time at a very different scale on the different hardwares, but other than that, it should be the same robot.

Let us now take this one step further. Instead of running the algorithm on conventional hardware, let us prepare a tape that specifies the state of the android and construct a Turing machine that implements its algorithm. This Turing machine is excruciatingly slow compared to the hyper-efficient hardware that was running the original copy of the android. However, in computational terms the Turing machine is completely equivalent to the original machine. There is no difference whatsoever, except in implementational terms. So, if the original android had an inner life, a FUDESS, now the Turing machine must have one as well.

We can now take this yet another step further. Instead of having a machine to do the state transitions, let us assume we write the transition rules for the reading head of the TM in a large table (similar to table 6.1 but much bigger). Then we pay an engineer to execute the transitions manually. We supply her with the input tape to the Turing machine and as many pencils and erasers she needs and let her do the transitions specified in the rule table. She would have to note down the state of the reading head, then read the symbol on the tape and consult the table to know what to do next. Depending on what it says, she would probably erase the symbol on the tape and write in a new one. Once she has done that she would then read the symbol to the left or right and consult the table again. And she would do this again and again and again. It would not be a very inspiring task for her, but we assume her to be meticulous and reliable so she would perform the computation correctly. Once we have reached this point, then there is not much left of our android and its alleged inner life. All she does is to manipulate meaningless symbols on a piece of paper. Yet, in computational terms this is the very essence of what is going on within the robot: the computation of the android's FUDESS, its inner lights. Whether the computation is performed within the metallic shiny humanoid body of the robot, or by a dusty engineer in a basement of a dilapidated building; whether it is done at lightening speed or over decades. Computation is computation. If the original android had an inner life, then the computational process that the engineer executes must also have an inner life, a FUDESS. Only, once the paraphernalia of sci-fi robotics are gone, the conscious robot does not seem to be so conscious any longer.

If the machine had a FUDESS, it would be very difficult to decide where it exists and when it exists. Just assume that our human engineer gets tired, and decides to call it a day after several hours of symbol manipulation. Does that mean that the android "dies" during the night, just to be resurrected the next morning when she comes back to work? What if we assumed that updating Turing machines is only a small part of her job, and she can only do one update per day. The computation would still continue, but now at an extremely slow speed. What about the subjective experience of the android? Does its FUDESS still exist? Does the android still experience its inner life, even though it is disembodied, and indeed dematerialised.

What if you are never shown the original android, but only

the Turing machine that is completely equivalent to it. How likely is it that you conclude from a complete understanding of its operation that it has a FUDESS. I would say it is rather unlikely. Instead of being instantiated in shiny metallic hardware, the android has now become a pen and paper being, a denizen of the Platonic universe of mathematics. Still, a computation is a computation. If it existed before and had an inner life, then it should continue to have an inner life, no matter where it is. Admittedly, the time-spans of the FUDESS are now enormous compared to the android in the original hardware. However, that should not matter. Computationally, it is all the same, whether the computation is slow or fast, it still is the same computation.

There are several conclusions to draw from this *gedankenexperiment*. The first one is that we are easily fooled by impressive behaviour and shiny user-interfaces into thinking that something is intelligent. Psychologically, we are quick to assign to a thing some inner life, just because it is able to hold up a conversation. I suspect, we are falling prey to our cognitive biases and conclude too quickly from good manners that there is a rich inner life. When we are interested in deep questions such as the possibility of an artificial FUDESS, then we must separate form from substance and avoid the cognitive bias.

This separation is quite important. As humans we are very prone to assume that all kinds of things have a FUDESS. My favourite example is the Italian cartoon "La Linea." On the spectrum of technical sophistications La Linea is at the extreme opposite of hyper-realistic cartoons such as Spielberg's Tintin. The level of detail, the realism of this cartoon is reduced to a minimum. All the viewer sees is a single line on a dark background. The line sketches the silhouette of something vaguely resembling a person. Not only is it very simple, but sometimes the viewer is even shown the hand of the person who draws the line. Add this to La Linea's extremely choleric character and you get a very engaging cartoon. Most people will buy into the concept eventually and regard La Linea as a "real" character. After some time, the viewer does not see a line any longer, but an irritable man in a 2d world who encounters all kinds of troubles on his way. Intellectually we know, of course, that it is just a drawing, but emotionally we are prepared to assign particular human qualities to a line. We feel that La Linea is an agent acting in its environment.

Even more so will we emotionally respond to autonomous robots that interact realistically with their environment. It does not take much to make us believe that they have an inner life, just as we are prone to grant La Linea an inner life. There seems to be a psychological button in us somewhere. When it is pressed we start to treat the opposite as a conscious agent. Evolutionarily, this is a good button to have. It would be very difficult to be a successful parent (and by implication a successful reproducer), if we were in doubt about the inner life of babies and whether or not they are more than mere objects. As long as the right kinds of cues are given, we are ready to accept anything as filled with a soul.

Coming back to our hypothetical android, let us assume that the algorithm that drives it is indeed sophisticated enough to generate a genuine FUDESS. I stress again, that this is a completely hypothetical assumption, because we can never be certain whether or not a FUDESS exists. The pen and paper plus engineer implementation of the algorithm would be completely equivalent to the full android. If the latter has a FUDESS so has the former. However, it does not press any of the psychological buttons and therefore it would also be very hard to convince anybody just by showing the transition rules and the input tape that there is an inner life in front of them. For any observer, the engineer would just manipulate symbols following some strict rules in a book with no apparent purpose.

Computers and their interfaces are designed so as to maximise the psychological impact they have on us. Realistic graphics in computer games can draw us into the action of the machine; a nice looking robot makes us empathise immediately.

Probably the optimism concerning naturalistic explanations of the mind-body problem is caused by researchers falling prey to the psychological effects of their own creations, the sophisticated interfaces and sexy branding. If we look at the evidence and just the evidence, then there is perhaps not so much reason for optimism. I do not remember ever having seen a truly convincing piece of artificial intelligence. There are computer programs that can do quite well at very specific tasks, but when it comes to a general type of intelligence they are disappointing. Much of the current optimism that FUDESS is going to be explained by scientific means alone is based on WFY arguments. Probably, and this is the suspicion I am trying to express here, the optimism is at least in part created by scientists falling for their own propaganda. They fall for the nice interfaces made to impress. When we take the artificial intelligence back to the Turing machine, much of the optimism would disappear. Moreover, much of the optimism is rooted in a belief that the problem of FUDESS is a technical problem. Get faster hardware, get better brain imaging and the problem solves itself. However, it seems that the main difficulty is a conceptual one. Even if we assume, in *gedankenexperiment*, that all the technical problems are solved, we always run into the Captain Kirk contradiction or related issues.

A preliminary conclusion we can draw from this is that the view of the brain as a computer and nothing but a computer does not explain much. Once we start to analyse what a computation fundamentally is, then we notice that any FUDESS that a computer may have disappears between our fingers. The process of computation reduces to the mechanical movements of the Turing machine.

Living computers and the freedom of will

Let us consider a different question now: Are we free? Free to make our own decisions? Free to choose what we want? Generations of philosophers have tried to solve this problem. So far, they have not reached a final conclusion. As it turns out, however, this very difficult question has a very simple solution if our brains are nothing but computers.

We all have the feeling that we are at least in part masters of our own deeds, our thinking and our acts. We can decide to do X and then actually do it. When we are faced with a decision, say, whether to listen to "Die Meistersinger von Nürnberg" or "Rheingold," then we can make a free choice. We could also decide to listen to "Tristan und Isolde." To some extent we are even masters of our own moods. When we are excited, we can sometimes calm ourselves down, or when we are tired we can excite ourselves. I suspect that good actors have the ability to get into particular states of mind on command so to speak. They can be sad in one moment and happy in another moment just as the scene requires.

There is no doubt that we cannot control ourselves completely. Some people, for example, have anger management problems. They cannot control eruptions of their tempers. Others may be permanently depressed and find no way out of their mood. Then again, there are now behavioural therapies that can teach people to control their inner states better. At least sometimes these techniques are successful. Another way to influence your inner states is to use meditation. Interestingly, Thomas Metzinger pleads for meditation as a potential way to change your state of mind for the better, maybe even to reach novel states of consciousness.

If we assume that the brain is just a big computer, then it is difficult to see how we can have any freedom to influence our own states of mind. Think again of a Turing machine. The actions of the Turing machine are completely deterministic. Once the transition rules and the input tape are given, there is no way in which the machine could depart from its pre-determined path.

Our android, that sits in a dark room and experiences its inner life, it would have no control over its thoughts. It probably felt like it was controlling what it is thinking, but it would not be. Maybe it decided to follow Metzinger's advice and practiced a bit of meditation, but this would just be an illusion. In reality, all that would have happened is that its program entered a "meditation" subroutine. That it would do so was pre-determined all along. The android did not have any control over that.

There are only two ways in which the paths of the Turing machine can be altered. Either one changes the input tape or the program that runs the Turing machine. By definition, a Turing machine cannot change its own program. It cannot change its input tape either. Hence, any perturbation away from the pre-determined trajectory of the Turing machine has to be external.

So, for example, after reading Thomas Metzinger's Ego Tunnel, the android may have become convinced that meditation is indeed a good way to learn more about consciousness, so it starts to practice. This is at least how it would have felt for the assumed android. In reality though, the book would have led to a reconfiguration of the input tape to the Turing machine such that meditation sub-routines are subsequently entered. If the brain is just a computer, then the same would be true for any other non-android reader of his book. If they feel they are convinced of his plea for meditation, then in reality they are not convinced, but they have been reprogrammed. It does not feel that way, but this is how it would have to be if we all were just computers.

What this suggests is that if we were just computers, then we would not be masters of our own thoughts, but mere passive observers of what goes on in our minds. Indeed, sometimes it does feel like this to all of us. We are probably all occasionally in the grip of compulsive thoughts that we cannot turn off. At other times, however, we feel much more in control of ourselves. Say, you are in a restaurant and decide what to order. Most of the time you will have an acute sense of freedom about what you will have on the menu.

How we feel about our decisions is irrelevant to the question of freedom. It is also irrelevant whether or not it seems that we make free decisions. Modern computer programs can for all practical purposes seem to be free. For some applications, such as artificial intelligence and trading algorithms this apparent freedom of choice is quite central to their purpose. However, a computer is unable to do anything that it has not been told to do, even if it seems otherwise. If we were computers, the very same argument would apply to us.

Revisiting SR

Denying FUDESS is often sold as the rational conclusion, the insight that heralds the beginning of a new understanding of humankind of itself. Typically, this is also taken as yet another proof that there is no God that created us. Metzinger is a case in point. He is convinced of the evidence against, not only the existence of Gods, but even the existence of FUDESS. He acknowledges that this may be unsettling and that some people may not be emotionally ready yet to come to terms with this conclusion. Still, he thinks that the evidence is so compelling that he calls for the "consciousness revolution." Like any good revolutionary, he sees enemies to the new world everywhere and feels the urgent need to protect the revolution "from all those who desperately seek emotional security and espouse closed world-views because they cannot bear the naturalistic turn in the image of humankind" (Noe, 2010). Quotes like this make me wonder how the new society will protect itself from the counter-revolutionaries. Will they be executed at the guillotine, or will they be sent into far away Gulags. What is it that Metzinger would do with all those who refuse to be enlightened, who want to remain in their pre-scientific closed world-views. Would we have to worry, in this new society that Metzinger envisions, that our neighbours would denounce us for believing in FUDESS?

Is Metzinger's case really as clear-cut as he likes to make us believe? Is every rational thinker compelled by scientific discoveries to come to the conclusion that there is no FUDESS? If so, is the evidence so strong that we can not only base our own *weltanschauung* on it, but that we also feel the need for a societal revolution? That we even feel we need to do something with those who cling to "closed world-views"? Is this an eternal undeniable truth that justifies an atheocratic organisation of society?

I do not think it is. The extremism that Metzinger displays is a perversion of any scientific ideal there is. The ethos of science is open-mindedness, not radicalism. Scientific insights no matter how certain they seem, are always preliminary. There are so many examples in science where widely held beliefs have been thrown out overnight. Scientific revolutions are rare, but frequent enough to be a real possibility. Hence, building a society and proposing to force-convert everybody to the new *weltanschauung* shows a complete misunderstanding of science. Unfortunately, Metzinger is not an isolated case. Intolerant and totalitarian tendencies are all too visible everywhere in the new fundamentalist atheism.

Science and the relative uncertainty of its conclusions by itself do not justify new world-orders. Yet, this fine point aside, what if we took scientific knowledge for granted for the time being. Based on the current state of the art of neuroscience, would it be rational to deny the existence of FUDESS and initiate the consciousness revolution that Metzinger is calling for? Are we compelled by research to deny the existence of FUDESS? Is there really a scientific case for denying the existence of the "me"?

The answer is yes and no. Yes, because science cannot deal with the "me." It is not a meaningful research question to ask about the subjective experience. It makes no sense to investigate how your perception of red differs from mine. We cannot find any language to compare subjective experiences, or even to formulate them. At the same time, we also have to answer "No." You, me, Thomas Metzinger and most others have intimate knowledge about their own experience. They know what Wienerschnitzel tastes like and what red looks like. They have knowledge of their own subjective experiences. This is meaningful knowledge. Probably even the only meaningful knowledge. The fact that this cannot be captured scientifically, says more about the limits of science, rather than about the existence of the self.

The confusion arises when we start to equate consciousness, as in consciousness research, with FUDESS. Clearly, consciousness research is not about subjective experience, it is about related things. If you believe SR, if you think that all there is can be studied with science, then of course, you must deny FUDESS, you must deny the self, the subjective experience, or you fall into the Captain Kirk contradiction. At the same time, a moment of thought will quickly show that science itself may be limited and SR is wrong. Yet, the proponents of SR, the science jihadists, cannot bear to let go of their SR closed world-view, the tenet that all valid knowledge is scientific knowledge. Hence, deny the existence of FUDESS. To me this makes it very clear that SR is not so much a rationally argued theory but a psychological condition.

Let us now look in more detail at the claim that the self/FUDESS does not exist. Metzinger goes to great lengths to make the claim...

> ...that the conscious experience of being a self emerges because of a large portion of the self-model in your brain is, as philosophers would say, transparent. We are Ego Machines, but we do not have selves. We cannot leave the Ego Tunnel, because there is nobody who could leave. The Ego and its Tunnel are representational. (Metzinger, 2009, pp. 207-8)

His thesis seems to be that the brain is a computational device, an internal modeller generating representations of the external world. So, probably Metzinger thinks of it in essence as a computational device that processes input to some suitable output. It is more than anything else a survival mechanism, an internal simulator that allows the animal to mentally play through scenarios before it takes any actions. Our feeling of autonomy is just a computational trick. Consciousness is a technique to run elaborate simulations, to play through scenarios mentally before they are realised in practice. So, according to Thomas Metzinger we are robots, biological computing machines, algorithm controlled zombies that roam the earth.

Is this position logically coherent? Can Thomas Metzinger, really claim that there are no selves, that there is no Thomas Metzinger, so to speak. If so, how can something that does not exist claim that it does not exist? To see this, imagine that Vishnu came down one day from his heavenly abode to reveal to us all that, in fact, we were wrong and he does not exist. It is a rather inconceivable event.

Before we get lost in speculations about the divine, let us take a step back and analyse the situation in more detail. Instead of Vishnu claiming his non-existence, let us assume a different scenario: Let us imagine programming a computer to print the words on screen, "I, nameless computer with IP address [fill in suitable number] do not exist." We can then run this program as often as we like and the computer will claim its own non-existence as often as we like it to do this. So, maybe it is not that absurd to claim your own non-existence when you are a computer. Clearly, though, the computer does not understand what it is printing and it can be equally programmed to claim its own existence. In fact, it can be programmed to print any conceivable statement. Whatever it prints it is inconsequential. So, trivially it is possible for something to claim that it does not exist, when it simply does not understand what it is claiming.

So what about Metzinger? If we follow Metzinger's own philosophy, then we have to conclude that he is, when push comes to shove, just a computer. A sophisticated one, but a computer nonetheless. He would be a carbon version of Commander Data, so to speak. This means that he is ultimately running some algorithm that determines what he is doing. Vastly more complex than the simple program that prints "I do not exist" on the screen, but still, just a program. Metzinger himself claims that nobody is at home in his body. He is indeed just a zombie, a machine that has amazing capabilities, but ultimately he just behaves as a mindless machine. If this is so, then he is of course absolutely right to claim his non-existence, as he implicitly does in his book. However, this would be coincidence. A little bit of re-programming, and he could equally have claimed his existence. Either way, it is inconsequential because there would be nobody to understand his utterances. He would just execute an algorithm. So, strictly speaking, any claims about himself can be safely disregarded.

Let us now consider the second case. Somebody is at home in Metzinger's body. He has motivations: probably he felt that he had something important to say, which motivated him to write a book in the first place. While he was typing his book, he actually felt the sensation of touching the keys with his fingers. On some days, he was probably very tired or he felt particularly excited about some of the chapters. Maybe, during one of his breaks from writing, he had a nice organic apple. When he bit into it he felt the cold juice oozing out of the fruit. If these sensations were indeed experienced by Metzinger, then it is fair to assume that the author of The Ego Tunnel had some first person experience and that his self indeed exists. The lights are switched on and there is a Metzinger FUDESS. If this is indeed so, then it is unclear to me how he could possibly claim that the self is non-existent without a very acute sense of inconsistence. When Metzinger says, "The self does not exist" which translates into "I do not exist," then he claims that there is nobody there to feel the apple, the keyboard, no one to be tired or excited.

Admittedly, I cannot know whether or not Metzinger exists and I have no way of finding out. He can only decide this for himself. However, I know for sure that I have a FUDESS as I am writing these words. Probably the reader has one as well. If you are in doubt, bite into an apple, or look around you, or listen to the sounds of your environment. If you can taste, or see, or hear anything, then you have a FUDESS. In this case, your brain is not merely a computational device that controls the body, but there is some experiencing entity beyond the computer.

For all of us who have a FUDESS, would it really be coherent to deny our own existence? Does it make sense to claim that there is no "I", no self when we experience it so clearly? Does it make sense for me to claim that I experience things, but at the same time I also claim that I do not exist? Hardly. In order to be able to formulate the thought and to utter the words "I do not exist" there has to be a thinking entity that makes this claim about itself. Whoever claims her/its own non-existence must be either speaking metaphorically, making an incoherent statement or does not understand what it is saying. I can conclude from this: Claiming non-existence of the self is logically incoherent for anybody who can feel the taste of an apple, experience the sensation of colours, or indeed experience anything at all.

There is a slightly milder claim, however: One may admit that the self exists and that there is somebody who experiences the sensation of biting into an apple. Yet, one may say, this self is causally impotent. It cannot do anything. It is a mere illusion of the brain, an *epiphenomenon* of the physical processes that go on in the brain. According to this idea the self is like a movie-goer in a hyper-realistic cinema. It experiences, it sees, it feels, not only what is going on in the movie, but also the inner states of one of the protagonist (herself) of the film. Yet, while she feels everything the protagonist is doing and saying, she cannot influence the plot. She is under the illusion of choosing her words and to think and act, but everything in the film has been determined by the film's director. She has no causal power to influence the film.

At a first glance epiphenomenalism is less problematic than a blank denial of the existence of the self. Certainly proponents of SR can now admit that there is a FUDESS without having to give up too much of the totality of science. All the important causal powers can still be studied as natural processes and they can be understood in terms of some laws of nature. The experience, the FUDESS, still remains somewhat mysterious, but it also becomes of secondary importance. It can be sidelined. It does not have any agency and is therefore of little practical importance. In any case, any philosophical threat from it for SR is banished. It is no longer a non-material, non-localisable command centre that somehow changes brain processes. Removing this unsavoury mind-matter interaction makes the idea of FUDESS much more palatable for proponents of SR. At a first glance it is very nice. We can now sit back, enjoy our illusion of FUDESS and most of all continue to do science without having to worry about this fundamental problem.

However, epiphenomenalism is not without its difficulties. Superficially it gets around the problem of FUDESS, in that it removes its causal influence on the material interactions of the brain. However, it is unclear how meaningful it is as a conjecture. One implication of epiphenomenalism is that there is no free will. Our states of minds, our desires, all our decisions are purely a function of the underlying mechanics of the brain which is the director of the film we are watching. Ultimately the neurons dictate what we are thinking, what we like and dislike. There is no way in which these felt and perceived states of mind can act back onto the states of the brain itself. As discussed above, the question of the freedom of will is ultimately not decided experimentally. There is no test one could perform to find out whether or not we are free to make our own decisions or not. Hence, a proponent of SR cannot hold this view without being inconsistent.

Consider the following scenario. Imagine that I am completely convinced that by my own free will and choosing I decided to do X or wanted to eat Y in the restaurant. Imagine, that this has been my wish for a long time and it came from my deepest inner me as a free choice. I could now go to an epiphenomenologist and try to tell her that she is wrong. I could try to argue that my personal experience of freely choosing the dish I am going to eat proves that there is free will. I just demonstrated it. I experienced my choice clearly and decided it is no illusion, it is real.

I am certain that the epiphenomenologist will be completely unimpressed by my argument. She will most certainly claim that my experience of free choice was but an illusion. In reality, I was just watching a program that was run by my brain. Maybe the epiphenomenologist will argue that my desire to eat dish Y rather than dish Z was itself not a free choice. I cannot wish to wish Z when in fact I wish Y. What appeared to be a free choice for me, was indeed a choice taken by the computer program running in my brain. This sounds convincing at first. However, the problem is that whatever argument for free will I propose to the epiphenomenologist, she will always be able to counter with her objection. Indeed, even if epiphenomenalism was wrong, she could still defend it. There is no possible argument I can make to convince the epiphenomenologist that her position is wrong. Another way to say this is that epiphenomenalism is self-immunising against criticism. There is no way to disprove it, in particular, there is no empirical test that could falsify it. Therefore, epiphenomenalism is more like the belief in Vishnu, rather than in Newton. It is not a scientific theory. Belief systems are precisely what proponents of SR detest most.

More fundamentally though, there are hidden conceptual problems with epiphenomenalism. The first one is of more philosophical nature. For the agent who thinks of herself as having a free will, it probably does not matter whether she really is free in some fundamental way or whether she just believes herself to be free. If in a restaurant, you have to choose what to order, it is irrelevant whether you just *believe* that you order what you want or whether your choice is genuinely free. The two possibilities are indistinguishable.

When the illusion of free will is indistinguishable from actual free will, then so is the illusion of the existence of FUDESS indistinguishable from actual existence. How can it in any way make a difference that the pain that you feel when somebody sticks a needle in your arm is just an illusion. Real pain hurts just as much as the perfect illusion of real pain. We cannot separate ourselves from the character of the hyper-realistic super-immersive cinema experience that epiphenomenalism tells us we live through. We may therefore just treat the illusion as reality. There is no difference as far as we are concerned. Philosophically, therefore, epiphenomenalism is an empty theory. It says nothing. Scientifically, it is empty as well. All it does is to define FUDESS out of science, or rather cover it up like an ugly patch of mould on a wall.

Digging just a little bit deeper, we soon encounter another problem: If free will and the conscious experience are just illusions, who is it that has the illusion. It is difficult to think of an illusion without also thinking of somebody who has the illusion. To be clear, we could think of material objects existing in isolation of any observer. Earth for example pre-dates any of its life forms. Hence, for a significant amount of time Earth was in existence, but was not observed by anybody. It makes perfect sense to assume this. Now, instead of thinking of Earth, imagine a mirage, the illusion of an oasis in a dry and hot dessert. We believe that the cause of a mirage is a particular pattern of light reflection on the hot surface of the dessert. These reflections surely exist without the existence of an observer. Still, the illusion of seeing an oasis in the middle of the barren dessert requires an observer. Somebody who not only detects the reflected rays of light, but attaches meaning to them, interprets them, or misinterprets them in this case. Or think of a drawing of a cube on a piece of paper. This is an illusion of a 3D shape on a two dimensional paper. In order to be elevated to an illusion from a mere drawing of lines it takes an observer who is deceived by it.

The same is true for FUDESS. If FUDESS is really just a mirage of the brain, then who is it that has this mirage? Who is tricked by the mind? Who has the illusion? How can an illusion be fooled by itself?

Chapter 7

Epilogue

> Zieht hin, zieht in den Berg der *Venus* ein!
> *Tannhäuser, Act 2*

Every age has to find its own answers to the fundamental mystery of existence, and the deep puzzle of why we exist. As we go about our daily business, it is very easy to take the world for granted. The moment we start to think about it we quickly realise that the existence of anything is a big mystery. That there is a universe rather than none is a contingent fact. It is even more surprising that there is a chain of transient existences that has led from the first unicellular proto-organisms to us. Then suddenly *you* are here. You have become matter that feels, matter that can feel pain and that has subjective experiences. This is not only puzzling, but also intellectually unsettling. What is it that feels, that suddenly makes the cold objective world light up for you? What is this subjective experience?

The universe is a bit like a stage in a cosmic play without an apparent audience. We are thrown in there in the middle of it all and we do not know whether we are the protagonists or the junior artistes. And we do not know whether we are appearing in a scripted play or whether it is all perhaps pure improvisation. Maybe the plot is made up as the play goes along.

Religion is a way to make sense of it all. The ancient Greeks embedded the world into a divine soap opera with anthropomorphic Gods that have their own human vanities and prides. Still above them is the impersonal fate, the providence that even the Gods cannot avoid. Within this framework of myths it was possible to anchor your own existence and to go on with your daily business without having to worry too much about why things *are* in the first place. Christians, Hindus and all the other faiths provide similar backstories for the existence of the universe. They also provide an audience for the cosmic play, the Gods.

Religious people are sometimes, at least implicitly, accused of being these weaklings, people who cannot come to terms with the reality of a meaningless universe and their own mortality. So they need to delude themselves by constructing myths of eternal life and supernatural beings that give hope and provide order. Maybe they are. If so, I would not blame them. Existence is scary. It is hard to see what is wrong with this desire for hope.

However, this accusation misses the point at least as a general critique of religions and their followers. It is a mistake to think that hope and promises of eternal life are universals of religion. The ancient Greeks are a case in point. Their shadowy underworld was an antithesis of hope. They had nothing whatsoever to look forward to after their earthly existences. Similarly, Judaism does not have a universal doctrine of life after death. Religion is not merely a way to delude yourself about your own mortality but satisfies some broader philosophical needs of humankind. It provides a cosmology, a way to make sense of the world, a tentative solution to the problem of the existence of the universe and of your own existence.

In many ways science answers the same needs. It provides a way to understand the world. People like Galilei and Monod were able to show that we can understand the rulebook of the universe. Not only is there meaningful knowledge to be found, but this knowledge can also be used to build devices and control the world around us. At least to some extent. This presses some of the buttons that religions press. In many ways prayers and rituals, pleading with the Gods has a similar function to technology. It enables, or seems to enable us to change things that would appear beyond our control otherwise.

In this sense, science and religion are on the same turf. Both can provide meaning and the feeling of exerting control where previously no control was possible. The problem is that a shared turf inevitably leads to turf wars. It is not surprising therefore, that science and religion are engaged in a culture war right now. This war is not about truth, it will not be decided based on who has the better arguments. Ultimately, culture wars of this sort will be decided based on needs. Which of the systems, natural science or religion, gives the better answers to our questions? Science seems to be slowly gaining the upper hand. It has a powerful narrative and it has a strong control over young peoples' minds as they go through schools.

Science provides a narrative about the world. Admittedly the narrative is about a meaningless world and at first it appears to be frightening and scary. On the other hand, it comes with much more. There are no shadowy underworlds in which the deceased have to reside. There are no devils or eternal punishments that await some. Science promises the big nothingness to everybody irrespective of class, academic title, income and faith. We all enter the Nirvana of total extinction. Yet, science comes with more. It also comes with hope and control. Scientists do not pray to any God to exert control, but they build things. Medical knowledge can cure diseases, engineering can protect us from the elements, science can help us avoid dangers.

The public purse funds science, not out of charity for the poor professors, but because science has a good story to tell. It captures the imagination and makes the sorts of promises that people like to hear. Of course, society expects something tangible in return for their money. Just as Jesus healed the lame and woke the dead, so does science back up its promises with deeds in the here and now. However, more often than not science does not keep its promises. Had scientists delivered on all the promises they made to get funding for their research projects, there would be no more disease and death in this world.

The core activity of science is not incompatible with religion.

Galilei's square law, Monod's insights, all these things do not even touch on the core messages of religions. The problem is that the backstory of science, the metaphysical superstructure, is bound to collide with the more established backstories from religion. Deliverance through science is not compatible with deliverance through Jesus. If the Nirvana of nothingness is true, then there cannot be the heavenly kingdom of Christianity or the seven heavens of Islam. There is only space for one gang on the religion turf.

In reality, the turf war is based on a misunderstanding. The scientific backstory of hope is not one that comes from science itself. It is a myth that scientists added over time to their science. This myth is itself not experimentally confirmed, it is not falsifiable and believing in it falls foul of any standard of scientific reasoning. Science does not deal with the afterlife or lack thereof. It cannot make any statements about the subjective experience or the FUDESS. Hence, it cannot tell us what happens after our death. Neither can it tell us where we were before we were born or indeed who we are. Still, there is the widespread impression that much of what science says seems to impinge on our concepts of the spiritual. Often it appears that scientific results show that there is no immortal soul, or that there is no God. After all, we would have found her by now. Yet, to think that science can say anything at all about the soul is a misunderstanding of science. Science does not do God. It simply cannot say anything about the spiritual, about what is outside the universe. Science must keep schtum about these things.

Science is about establishing facts, predicting experiments, constructing theories about the world. Quality science always has to remain anchored in the empirical. The moment science makes claims that have no chance of ever being tested experimentally, it leaves the realm of science and enters the parts of the intellectual universe inhabited by witchcraft, astrology and shamanism. Rational thinkers are not entitled to conclude from facts about the material world metaphysical questions about the universe. It is not possible to prove or disprove the existence of Gods from an understanding of the genetic regulatory networks of *E.coli*, or from any fact of neuroscience.

A commonly held misconception is the following:

Evolution is true, hence no God exists.

I have never found this story particularly convincing, but I suspect it makes more sense for those who make naive and unimaginative assumptions about Gods. Evolution is perhaps incompatible with religion if you think that God is an old man with long white beard who created Adam and Eve out of clay. Instead we could think of God as a computer modeller in a higher order universe, just like Maxime C. If God was more like this, then suddenly the contradiction between evolution and science disappears. If I were God, I would probably create the world through evolution. Maybe I could come up with a better way, but I would not dirty my hands with clay.

Many more similar fallacies litter popular scientific literature. A particularly powerful one is:

Consciousness is a physical process and hence there exists no immortal soul.

As an argument this is much more intelligent than the above evolution point, but ultimately there is a dodgy implicit assumption in this along with a fallacy. The dodgy assumption is the WFY argument that is hidden here. As a matter of fact we understand very little about consciousness and the brain as of now. Whether we really will get to a position where we have a full theory of the mind is unknown now. It may or may not happen. It is impossible to predict this with any degree of confidence. Still, even if we have a full theory of mind, science would still have to keep schtum about the metaphysical implications of any such theory. A physical theory of the mind cannot explain the subjective experience, no matter how advanced it is. The Captain Kirk contradiction is looming here and remains unsolvable by science alone—on principle grounds, not because of a technological limitation, or a limitation of knowledge. I think the only honest conclusion to draw is that we do not understand this yet. Our conceptual tools are at present not sophisticated enough to understand FUDESS and its relationship to matter. I suspect that the main problem is the flat-packing. Science is very good at uncovering the mechanistic basis of phenomena. This is also tremendously satisfying intellectually, but probably not the right approach to understand complex systems as a whole. When the Galileian approach does not work, then scientific explanations are simply not as good any more.

Related to this is a third fallacy that is often encountered. It is the neuro-recductionism fallacy. It goes as follows:

The desire to believe in a God is built into our brains [description of some experiments follow], hence God is just imagined [which is tacitly taken to mean that she does not exist].

This is one of the more naive ones, of course. Independent of the existence of Gods, we certainly need to have the cognitive abilities to believe in them and a desire or an inclination to do so. It does not need a detailed ethnographic or sociological study to show that we have both. This implies that this desire must be somehow built into our brains. It must also have provided us with some sort of adaptive benefit during our evolution, at least it must not have been too detrimental, otherwise we would have died out. In any case, if we were not hard-wired for belief, then we would not believe. However, the hard wiring does not say anything about the existence or non-existence or the nature of Gods. It is simply a statement about our neuro-biology.

Any scientist has to turn into an atheist during work hours. The assumption of Gods simply does not make any sense within science. The moment a researcher starts to invoke the divine in her work, she will do very bad science. A professional boxer has to forget about his sympathies for his opponent when he is in the ring. The fact that scientists have to exclude Gods during their work does not prove that there is no God and the fact that the boxers fight one another in the ring does not show that they actually hate one another. The practical atheism of a scientist is a bit like the faith of a Formula 1 driver in his car. If he worried about the tyres bursting or that the steering column will break, then he would be a miserable driver.

Science is a social activity that fulfils important functions in modern societies. There are others. There is no reason to assume that science is privileged amongst those activities and there is certainly no reason to assume that its practical atheism can be extended to a general societal principle, just as the opposing boxers may not dislike each other outside the ring.

The illusion of control

It is a commonplace to say that science has enormous impact on society, technology and engineering, modern medicine. Our lives would be very different—definitely worse—without science. Often science seems like a miracle. Just think of a telephone. We are used to it, but when you think about it for a moment, you will see just how extra-ordinary it is to be able to talk to somebody who is at the other side of the world. For someone who has never seen a telephone, it must seem like magic. The telephone is not the only science miracle. There are many more. Indeed, our daily lives are full of these little miracles. Every single phone call, each time you turn on your computer and whenever you use a piece of technology or visit your doctor, in all of those cases you experience the power of science.

It is easy to be impressed by this and to forget the limitations of science. All the things science does not know. The knowledge that we have not gained. Yet, a closer look at science and how it really operates makes it plain how little we know and how strongly science is limited by its own methods. Science works very well when it can idealise. Whenever it is possible to isolate things in laboratories, then the research results are solid. Galilei did it, Monod did it and many others did it.

It is surprising just how far the Galileian paradigm gets us. At the same time, powerful as it is, the Galileian approach is also limited. Once science goes beyond these limits it becomes an entirely different thing. To some extent this is already apparent in the Monodian approach. Even though Monod still keeps up appearances, some of the main features of Galieli have already been sacrificed. There is no theory to speak of in biological sciences. Research goes from hypothesis to hypothesis, but there is no general framework that allows all the particular insights to be understood as special cases of a more general theory. Biology is about particulars.

Once science goes out of the lab things become even harder. Science about the real complex world is chronically unreliable. It is a far cry from the neatness of Galilei. When it comes to complex systems, scientific prediction can be worse than random guessing. It can certainly be more dangerous. In the lab the scientist has developed the illusion of knowledge and control. Only once she leaves the lab, all the control disappears with the ability to build laboratory systems. The environment is no longer isolated. Anything can happen. This makes economics so hard and its predictions so unreliable. One wonders why researchers still bother to make them and why the politicians still bother to listen.

I suspect that this is largely a psychological phenomenon. Society has a need to feel in control and it also has a belief in science. Hence, it is comforting to believe that economists still have something to say, although past experience should have taught us otherwise. In a way, with economics it is all not so tragic and they are taken less and less seriously. Things get worse when it comes to predictions about large scale technologies. Physical scientists still carry a lot of clout. They are trusted by the public and they are very sure of themselves. Only, they tend to delude themselves about the uncertainties of their own predictions, just as the economists delude themselves. When they are trusted but get it wrong, the consequences can be serious. The climate is destroyed or a nuclear power station melts after being hit by a tsunami.

The illusion of control is dangerous, but it is also intellectually misleading. We are all too willing to forget about the limitations of science and fall prey to the evangelists of SR. They tell us that science is all-powerful, that we should organise our society according to scientific principles. They teach the gospel of scientific rationalism and they want us to join the consciousness revolution. Naturally, one is inclined to believe them because science has performed so many miracles.

A balanced view

There is a culture war going on out there. On the one hand there are the religious fundamentalists who cling to their beliefs. They feel the dominance of science intruding on their way of life. And they feel threatened by it. Some people define themselves through their scientific outlook on life. Their rationalism is who they are. Then there are people who define themselves through their religious beliefs. Some of those may feel that their mental wellbeing depends on the set of dogmas they are adhering to. In western societies the most common incarnation of this type of person is the Christian fundamentalist.

Those fundamentalists get into conflict with science. When they feel compelled to take the message of Genesis as a factual account of what happened at the beginning of the world, then it is hard for them to combine this account with Darwin's gradual evolution or the big bang hypothesis. If you are one of those, then you are faced with either of two possibilities. You can make sure you deny the science,or you can change your view on religion. If you have invested significant social and psychological effort in being a fundamentalist Christian then you will more likely turn into one of those people who decries the teaching of evolution in schools; yet some turn straight around and become evangelical atheists overnight.

Fundamentalist Christianity is a hopeless program. Attempts to reconcile a literal view of the bible with science nearly always result in rather embarrassing pseudo-scientific arguments. It is not even worth discussing those arguments, a point of view that is incidentally shared by the more honest fundamentalists who refuse rational exchange, but go straight into science denial mode. Instead of arguing against science, they just ignore it.

When it comes to science and religion though, this effortful denial is unnecessary. The bible cannot and should not be taken literally. For one, being the product of human authors and written over a long period of time, the Christian scriptures are full of internal contradictions, inconsistencies and errors. Consequently the fundamentalists have to spend significant amounts of mental energy to explain them away or to simply not think about them. A sober assessment of the situation tells us straight away that not everything in it can be true. Even if there were none of those glaring mistakes in the bible, it would still be quite impossible to understand the scripture literally, as it is indeed impossible to take any text with a moderately complex content literally. Natural language is not suited for unambiguous communications. Most of the time, it is not necessary either to be unambiguous. It is sufficient to get a rough message across; common sense and cultural conventions then remove most of the ambiguity. With all this in place communication is then sufficiently clear. In the rare cases where precision is of the essence, texts can get very difficult to understand. An example is law texts that can be quite impenetrable for the uninitiated. Sometimes natural language is not sufficient any longer and new languages have to be invented. Mathematics is another example. It has its own syntax and conventions that make it possible to be precise and unambiguous to a high degree.

The bible, particularly the New Testament, is not intended as a precise text in this sense. Particularly the gospels rely in large parts on parables to get their point across. How can this possibly be taken literally? There is no way of understanding the bible without interpreting it. It is intrinsic to interpretations that they are subjective to the interpreter, to some extent at least. Some interpretations are more reasonable than others. Still, for any given text there is a spectrum of possibilities that cannot be distinguished. All are equally true.

The question is now why so many of the fundamentalist Christians even feel that there is a need for a literal interpretation of scripture? To me it seems that the insistence on literal truth is yet another symptom of our scientific *weltanschauung*. The irony is that fundamentalists Christians read the bible as contemporaries of a scientific age. They have forgotten how to interpret scriptures, so we read them as factual accounts of past events. The irony is that fundamentalists deny science, fight evolution because they themselves see the world through the scientific spectacles.

One does not need to look far to see the battles of the culture war that is going on. Science against religion, rationalism against belief, proper medicine against alternative healing and so on. I agree that there are some questions to be debated. As a society we constantly have to recalibrate our *weltanschauung*. Just as all the generations before us, we have to find our own answers to the deep mysteries of existence, and we have to find ways to integrate them into our practices. Somehow we have to find a way to understand the world around us and to make sense of the here and now. We also need to think about ways to organise society and how this can reflect our understanding of the world. All this involves metaphysics, science and religion. There is a need to have a debate between various aspects of our culture.

However, there is no need for a culture war. It is just as silly to be anti-science as it is to blindly follow SR. It is equally silly to believe that religion is a bad thing, or worthless, just because it is not scientific. There is more to life than science. Fighting religion is also futile. Along with prostitution, religion is one of the great human universals that cannot be eradicated, no matter how hard you try. It is a need we have within ourselves. Some have more of this need, some less, but generally religion will stay. The form of religion might change, but religion in some form cannot be eradicated. Indeed, this is confirmed by the evangelical atheists who religiously fight religion. They themselves have erected belief systems that are not so different from those of the evangelical Christians they so detest.

Bibliography

1 J. Monod. The growth of bacterial cultures. *Annual Review in Microbiology*, 3:371–349, 1949.

2 A. Pardee, F. Jacob, and J. Monod. The genetic control and cytoplasmic expression of "inducibility" in the synthesis of β-galactosidase by *E. coli*. *Journal of Molecular Biology*, 1:165–178, 1959.

3 Monica Riley, Arthur B. Pardee, Francois Jacob, and Jacques Monod. On the expression of a structural gene. *Journal of Molecular Biology*, 2(4):216 – 225, 1960.

4 F. Jacob and J. Monod. Genetic regulatory mechanisms in the synthesis of proteins. *Journal of Molecular Biology*, 3:318 – 356, 1960.

5 J. Elf, G. Li, and X. Xie. Probing transcription factor dynamics at the single-molecule level in a living cell. *Science*, 316(5828):1191–1194, May 2007.

6 J. Epstein and R. Axtell. *Growing Artificial Societies*. The MIT Press, Cambridge, 1998.

7 U. Pesavento. An implementation of von neumann's self-reproducing machine. *Artifial Life*, 2(4):337–354, 1995.

8 J. Stricker, S. Cookson, M. Bennett, W. Mather, L. Tsimring, and J. Hasty. A fast, robust and tunable synthetic gene oscillator. *Nature*, 456(7221):516–519, Nov 2008.

9 J. Donaldson and A. Scheffler. *Monkey Puzzle*. Macmillan Children's Books, 2000.

10 C. Cook. Sangomas: problem or solution for south Africa's health care system? *Journal of the National Medical Association*, 101(3):261–265, Mar 2009.

11 T. Kaptchuk. The placebo effect in alternative medicine: can the performance of a healing ritual have clinical significance? *Annals of Internal Medicine*, 136(11):817–825, Jun 2002.

12 F. Miller and T. Kaptchuk. The power of context: reconceptu-

alizing the placebo effect. *Journal of the Royal Society of Medicine*, 101(5):222–225, May 2008.

13 L. Colloca, L. Lopiano, M. Lanotte, and F. Benedetti. Overt versus covert treatment for pain, anxiety, and Parkinson's disease. *The Lancet Neurology*, 3(11):679–684, Nov 2004.

14 T. Mayer. The roman inquisition's precept to Galileo (1616). *The British Journal for the History of Science*, 43(03):327–351, 2010.

15 Thomas Metzinger. *The Ego Tunnel: The Science of the Mind and the Myth of the Self*. Basic Books, 1 edition, March 2009.

16 David J. Chalmers. Subsymbolic computation and the Chinese room. In *The Symbolic and Connectionist Paradigms: Closing the Gap*. Lawrence Erlbaum, 1992.

17 Alva Noe. Out of our heads: Why you are not your brain, and other lessons from the biology of consciousness. *Minds Mach.*, 20:155–159, February 2010.

18 R. Fjelland. Vitenskap mellom sikkerhet og usikkerhet. Ad Notam Gyldendal, Oslo, 1999.

19 R. Fjelland. The Theory-Ladenness of Observations, the Role of Scientific Instruments, and The Kantian a priori. International Studies in the Philosophy of Science, 5(3):269–280, 1999. 165

20 R. Rosen. Life Itself. Columbia University Press, New York, 1991.

21 R. Rosen. Essays on Life Itself. Columbia University Press, New York, 1999.

22 A. Pickering. The Mangle of Practice. The University of Chicago Press, Chicago, 1995.

23 A. Pickering. Constructing Quarks. The University of Chicago Press, Chicago, 1984.

24 K. Popper. The Logic of Scientific Discovery. Hutchinson, London, 1959.

25 R. Giere. Understanding Scientific Reasoning. Holt, Rinehart and Winston, Chicago, 1984.

26 R. Strand. The Role of Risk Assesment in the Governance of Genetically Modified Organisms in Agriculture. Journal of Hazardous Materials, 86:187-204, 2001.

27 R. Strand. Towards a Useful Philosophy of Biochemistry: Sketches and Examples. Foundations of Chemistry, 1:271–294, 1999.

Endnotes

1 There will always be some inaccuracies due to various factors, including inaccuracies to determine the precise time the last grain has fallen. But this is a technical difficulty that we will ignore for the moment.

2 I do acknowledge that the assumption that our protagonists live in a Newtonian world may stretch the reader's imaginative abilities.

3 The following quotations were taken from the MIT Open Courseware homepage: http://ocw.mit.edu/courses/physics/8-01-physics-i-classical-mechanics-fall-1999/video-lectures/lecture-6/.

4 Although, the types of experiments that are permissible to perform will depend on values. Particularly in medical research there are strict, value-based boundaries on what can and what cannot be done. Experiments with human embryos, for example, are subject to strict controls to prevent unethical behaviour.

5 http://climateprediction.net

6 This is strictly speaking an assumption on my part. I can only be certain about my own first person experience. It may well be that no one except myself has it, even though everybody acts as though they do.

7 http://jahia-prod.epfl.ch/site/bluebrain/

BOOKS

Iff Books is interested in ideas and reasoning. It publishes material on science, philosophy and law. Iff Books aims to work with authors and titles that augment our understanding of the human condition, society and civilisation, and the world or universe in which we live.